Kelley Wingate

Pre-Algebra

Grades 5–8

Credits
Editor: Katie Kee
Copy Editor: Beatrice Allen

Visit *carsondellosa.com* for correlations to Common Core, state, national, and Canadian provincial standards.

PO Box 35665
Greensboro, NC 27425 USA
carsondellosa.com

ISBN 978-1-4838-0504-7
02-097141151

Table of Contents

Introduction

Competency in math skills creates a foundation for the successful use of math principles in the real world. Practicing pre-algebra skills is the best way to improve at them.

This book was developed to help students practice and master a variety of pre-algebraic concepts. The practice pages can be used first to assess proficiency and later as basic skill practice. The extra practice will help students advance to more challenging math work with confidence. Help students catch up, stay up, and move ahead.

Common Core State Standards (CCSS) Alignment

This book supports standards-based instruction and is aligned to the CCSS. The standards are listed at the top of each page for easy reference. To help you meet instructional, remediation, and individualization goals, consult the Common Core State Standards alignment chart on page 4.

Leveled Activities

Instructional levels in this book vary. Each area of the book offers multilevel math activities so that learning can progress naturally. There are three levels, signified by one, two, or three dots at the bottom of the page:

- Level I: These activities will offer the most support.
- Level II: Some supportive measures are built in.
- Level III: Students will understand the concepts and be able to work independently.

All children learn at their own rate. Use your own judgment for introducing concepts to children when developmentally appropriate.

Hands-On Learning

Review is an important part of learning. It helps to ensure that skills are not only covered but are internalized. The flash cards at the back of this book will offer endless opportunities for review. Use them for basic skill and enrichment activities to reinforce basic pre-algebraic concepts.

There is also a certificate template at the back of this book for use as students excel at daily assignments or when they finish a unit.

Common Core State Standards Alignment Chart

Common Core State Standards*		Practice Page(s)
Operations and Algebraic Thinking		
Write and interpret numerical expressions.	5.OA.1, 5.OA.2	62–64, 83–85
Number and Operations in Base Ten		
Perform operations with multi-digit whole numbers and with decimals to hundredths.	5.NBT.7	8–22
Geometry		
Graph points on the coordinate plane to solve real-world and mathematical problems.	5.G.1	95–97
Ratios and Proportional Relationships		
Understand ratio concepts and use ratio reasoning to solve problems.	6.RP.1–6.RP.3	26–28, 32–34
Analyze proportional relationships and use them to solve real-world and mathematical problems.	7.RP.1–7.RP.3	29–34, 38–40
The Number System		
Apply and extend previous understandings of multiplication and division to divide fractions by fractions.	6.NS.1	5–8
Compute fluently with multi-digit numbers and find common factors and multiples.	6.NS.3	8–22
Apply and extend previous understandings of numbers to the system of rational numbers.	6.NS.5–6.NS.7	41–46, 95–97
Apply and extend previous understandings of operations with fractions to add, subtract, multiply, and divide rational numbers.	7.NS.1, 7.NS.2	23–25, 35–37, 47–58
Expressions and Equations		
Apply and extend previous understandings of arithmetic to algebraic expressions.	6.EE.2, 6.EE.3	62–67, 86–88, 92–94
Use properties of operations to generate equivalent expressions.	7.EE.1	59–61
Solve real-life and mathematical problems using numerical and algebraic expressions and equations.	7.EE.3, 7.EE.4	68–85, 89–91
Understand the connections between proportional relationships, lines, and linear equations.	8.EE.5	101–103
Analyze and solve linear equations and pairs of simultaneous linear equations.	8.EE.7	98–100
Functions		
Use functions to model relationships between quantities.	8.F.4	101–103

Name ______________________________ 6.NS.A.1

Dividing Fractions

$$1\frac{2}{3} \div 2\frac{1}{5} = \frac{5}{3} \div \frac{11}{5} = \frac{5}{3} \times \frac{5}{11} = \frac{25}{33}$$

rewrite — invert and multiply — rewrite

Solve each problem. Write the answer in simplest form.

1. $6\frac{2}{3} \div 4\frac{4}{9} =$

2. $3\frac{1}{3} \div 1\frac{5}{9} =$

3. $2\frac{7}{10} \div 3\frac{9}{15} =$

4. $4\frac{1}{2} \div 5\frac{1}{4} =$

5. $6\frac{3}{4} \div 2\frac{1}{2} =$

6. $2\frac{2}{6} \div 4\frac{2}{3} =$

7. $5\frac{2}{5} \div 4\frac{1}{2} =$

8. $7\frac{2}{7} \div 2\frac{2}{14} =$

9. $3\frac{1}{2} \div 4\frac{1}{3} =$

10. $2\frac{2}{3} \div 3\frac{4}{10} =$

11. $4\frac{1}{5} \div 3\frac{3}{5} =$

12. $5\frac{3}{5} \div 1\frac{5}{9} =$

Name ________________________________ 6.NS.A.1

Dividing Fractions

rewrite · invert and multiply

$$1\frac{1}{8} \div 2\frac{1}{6} = \frac{9}{8} \div \frac{13}{6} = \frac{9}{8} \times \frac{6}{13} = \frac{27}{52}$$

rewrite

Solve each problem. Write the answer in simplest form.

1. $9\frac{1}{6} \div 3\frac{5}{12} =$
2. $9\frac{1}{6} \div 3\frac{8}{12} =$
3. $7\frac{1}{2} \div 8\frac{3}{4} =$
4. $5\frac{1}{2} \div 8\frac{4}{5} =$
5. $5\frac{4}{5} \div 1\frac{8}{15} =$
6. $9\frac{1}{5} \div 2\frac{3}{10} =$
7. $7\frac{4}{5} \div 1\frac{3}{10} =$
8. $7\frac{1}{9} \div 2\frac{2}{3} =$
9. $8\frac{4}{5} \div 1\frac{1}{15} =$
10. $8\frac{2}{5} \div 2\frac{1}{10} =$
11. $5\frac{3}{5} \div 1\frac{6}{10} =$
12. $6\frac{1}{3} \div 2\frac{1}{6} =$
13. $11\frac{3}{4} \div 5\frac{1}{2} =$
14. $8\frac{3}{5} \div 2\frac{7}{10} =$
15. $3\frac{5}{7} \div 3\frac{13}{14} =$

Name ______________________________ 6.NS.A.1

Dividing Fractions

Solve each problem. Write the answer in the simplest form.

1. $\frac{1}{4} \div 4 =$

2. $1\frac{1}{6} \div 2\frac{1}{2} =$

3. $\frac{5}{6} \div \frac{2}{3} =$

4. $\frac{1}{5} \div \frac{1}{6} =$

5. $\frac{3}{5} \div \frac{7}{10} =$

6. $2\frac{4}{7} \div \frac{1}{8} =$

7. $3\frac{1}{3} \div 5\frac{1}{2} =$

8. $3\frac{1}{5} \div 1\frac{6}{10} =$

9. $2\frac{2}{9} \div 4\frac{1}{6} =$

10. $4\frac{3}{5} \div 1\frac{3}{8} =$

11. $3\frac{3}{4} \div 3\frac{1}{8} =$

12. $9\frac{3}{7} \div 5\frac{10}{14} =$

13. $5\frac{1}{6} \div 2\frac{7}{12} =$

14. $\frac{8}{7} \div 4 =$

15. $1\frac{1}{2} \div 2\frac{1}{4} =$

16. $6\frac{1}{5} \div 2\frac{2}{3} =$

17. $1\frac{2}{3} \div 1\frac{1}{3} =$

18. $1\frac{2}{5} \div 2\frac{3}{4} =$

19. $3\frac{1}{9} \div 12\frac{2}{8} =$

20. $3\frac{4}{10} \div 3\frac{6}{7} =$

21. $4\frac{5}{9} \div 3\frac{5}{12} =$

Name ____________________ 5.NBT.B.7, 6.NS.B.3

Adding Decimals

$$12.2 + 5.25 = \begin{array}{r} 12.20 \\ +\ 5.25 \\ \hline 17.45 \end{array}$$

Solve each problem.

1. 9.87 + 2.87 =
2. 5.02 + 8.2 =
3. 2.49 + 4.73 =
4. 6.41 + 2.734 + 8.41 =
5. 2.934 + 231.6 =
6. 121.9 + 0.736 =
7. 43.56 + 85.7 =
8. 13.238 + 4.82 =
9. 15.76 + 25.23 + 3.9 =
10. 6.41 + 3.99 =
11. 6.3 + 9.124 + 2.34 =
12. 5.97 + 4.87 + 3.908 =
13. 13.39 + 7.4 =
14. 4.63 + 23.5 + 5.0 =
15. 3.456 + 2.894 =
16. 3.64 + 5.32 =
17. 5.7 + 5.34 + 4.78 =
18. 3.5 + 8.4 =
19. 0.034 + 10.51 =
20. 123.415 + 6.876 =

Name ______________________

5.NBT.B.7, 6.NS.B.3

Adding Decimals

Remember:
To add decimals, align the numbers on the decimal. Add zeros as place holders. Then, add.

Solve each problem.

1. 40.14 + 12.53 + 5.6 =
2. 3.43 + 5.45 =
3. 3.59 + 2.08 =
4. 17.34 + 6.45 =
5. 2.15 + 4.25 =
6. 108.7 + 0.489 =
7. 31.71 + 324.95 =
8. 121.356 + 80.52 =
9. 5.37 + 7.37 =
10. 7.22 + 3.41 =
11. 2.6 + 45.54 + 3.65 =
12. 6.29 + 8.83 + 6.332 =
13. 5.4 + 7.38 + 6.21 =
14. 8.45 + 23.20 + 5.34 =
15. 5.44 + 3.34 + 6.30 =
16. 2.312 + 5.371 =
17. 12.52 + 8.32 =
18. 321.595 + 3.45 =
19. 0.012 + 25.08 =
20. 17.121 + 5.34 =

Name ____________________ 5.NBT.B.7, 6.NS.B.3

Adding Decimals

Solve each problem.

1. 42.25 + 53.5 =

2. 12.828 + 10.548 =

3. 0.6 + 0.09 + 1.75 =

4. 65.78 + 54.90 =

5. 6.77 + 0.05 =

6. 20.59 + 44.5 =

7. 154.34 + 42.98 =

8. 17.546 + 5.0958 =

9. 23.565 + 28.403 =

10. 0.8 + 0.07 + 3.73 =

11. 4.35 + 5.4 =

12. 52.4 + 0.62 + 2.096 =

13. 16.08 + 3.2 =

14. 22.3 + 0.89 + 27 =

15. 7.150 + 0.263 =

16. 0.258 + 32.900 =

17. 75.10 + 3.62 =

18. 13.80 + 6.32 =

19. 0.894 + 1.773 =

20. 12.354 + 15.438 =

Name ____________________

5.NBT.B.7, 6.NS.B.3

Subtracting Decimals

$$13.2 - 4.10 = \begin{array}{r} 13.20 \\ -\ 4.10 \\ \hline 9.10 \end{array}$$

Solve each problem.

1. 4.239 − 0.06 =

2. 51.23 − 14.45 =

3. 16.3 − 12.4 =

4. 452.82 − 127.36 =

5. 62.1 − 33.29 =

6. 75.034 − 22.439 =

7. 76.34 − 47.30 =

8. 34.32 − 12.43 =

9. 435.34 − 345.34 =

10. 756.98 − 32.43 =

11. 513.43 − 305.342 =

12. 65.9 − 33.2 =

13. 21.73 − 16.43 =

14. 121.32 − 19.34 =

15. 23.28 − 0.552 − 1.2 =

16. 8.64 − 0.476 =

17. 13.2 − 6.7 =

18. 21.32 − 4.28 =

19. 35.63 − 0.021 =

20. 485.02 − 332.86 =

Name ____________________ (5.NBT.B.7, 6.NS.B.3)

Subtracting Decimals

Remember:
To subtract decimals, align the numbers on the decimal. Add zeros as place holders. Then, subtract.

Solve each problem.

1. 43,289.56 − 28,125.87 =
2. 756.84 − 31.343 =
3. 34.34 − 23.19 =
4. 4.7 − 2.3 =
5. 95.87 − 52.45 =
6. 72.72 − 43.562 =
7. 85.76 − 34.65 =
8. 7.435 − 0.0345 =
9. 345.24 − 159.24 =
10. 54.68 − 23.76 =
11. 74.71 − 61.92 =
12. 84.8 − 44.87 =
13. 857.44 − 22.39 =
14. 93.76 − 8.67 =
15. 233.23 − 6.45 =
16. 6.56 − 0.654 =
17. 43.5 − 0.015 − 3.2 =
18. 39.43 − 15.34 =
19. 56.4 − 0.043 =
20. 954.34 − 657.56 =

Name ______________________________

5.NBT.B.7, 6.NS.B.3

Subtracting Decimals

Solve each problem.

1. 725.987 − 231.155 =
2. 13.58 − 7.2 =
3. 432.42 − 327.89 =
4. 1,343.32 − 1,032.90 =
5. 21.7 − 15.9 =
6. 843.21 − 452.03 =
7. 15.51 − 8.34 =
8. 545.825 − 137.405 =
9. 43.5 − 14.2 =
10. 1,350.65 − 253.42 =
11. 2.55 − 2.05 =
12. 5.085 − 2.5 =
13. 12 − 2.73 =
14. 26.8 − 6.48 =
15. 2.04 − 0.998 =
16. 66.5 − 0.675 =
17. 35.91 − 17.16 =
18. 0.9505 − 0.02709 =
19. 1 − 0.98765 =
20. 2.35 − 0.9 =

Name ____________________ 5.NBT.B.7, 6.NS.B.3

Multiplying Decimals

(0.3)(0.12) — 3 decimal places → 0.036

$$\begin{array}{r} 0.3 \\ \times\ 0.12 \\ \hline 0.036 \end{array}$$

Solve each problem. Show your work.

1. (0.6)(0.022) =
2. (0.012)(0.7) =
3. (3.2)(0.65) =
4. 0.07 × 0.4 =
5. (0.02)(1.2) =
6. 0.03 × 0.7 =
7. (0.5)(0.2) =
8. (2.2)(0.22) =
9. (0.12)(0.04) =
10. 0.06 × 0.07 =
11. (0.11)(0.07) =
12. (0.13)(0.02) =
13. (0.7)(0.07) =
14. (0.5)(0.05) =
15. 0.5 × 0.06 =
16. (0.012)(1.2) =
17. (0.8)(0.005) =
18. 0.25 × 0.07 =
19. (0.9)(0.002) =
20. (0.9)(0.9) =

Name ____________________ 5.NBT.B.7, 6.NS.B.3

Multiplying Decimals

Remember:
To multiply decimals, align the numbers on the right side. Then, multiply. Place the decimal point in the product so it has the same number of decimal places as the factors.

Solve each problem. Show your work.

1. (7.8)(1.03) =

2. (3.2)(3.065) =

3. (12.2)(34.9) =

4. (0.04)(0.24)(1.4) =

5. (2.5)(3.3)(0.33) =

6. (1.3)(3.04)(5.46) =

7. (4.3)(3.59) =

8. (5.5)(4.304) =

9. (23.4)(3.9) =

10. (5.5)(2.6)(4.0) =

11. (7)(20.2) =

12. (6.2)(0.35) =

13. (0.2)(0.18) =

14. (8.5)(9.1) =

15. (4.1)(4.1) =

16. (5.014)(5.4) =

17. (7.5)(0.75) =

18. (5.75)(1.2) =

19. (65.7)(2.5) =

20. (8.2)(0.2) =

Name ____________________

5.NBT.B.7, 6.NS.B.3

Multiplying Decimals

Solve each problem. Show your work.

1. (12.3)(5.81)(0.06) =

2. (0.042)(0.006) =

3. (0.34)(0.12)(0.104) =

4. (8.9)(0.11)(3.09) =

5. (15.92)(0.4)(0.32) =

6. (0.004)(6) =

7. (6.4)(0.3) =

8. (0.4)(0.232) =

9. (5.12)(6) =

10. (10.89)(0.221) =

11. (3.28)(12.8) =

12. (0.004)(0.0004)(0.04) =

13. (0.016)(3.8) =

14. (0.007)(0.6)(0.05) =

15. (3.806)(10.01) =

16. (340)(0.02) =

17. (0.8)(0.342)(0.02) =

18. (2.09)(0.005) =

19. (0.05)(0.15)(0.002) =

20. (5.4)(0.645)(0.07) =

Name ______________________________

5.NBT.B.7, 6.NS.B.3

Dividing Decimals

$1.2\overline{)0.84}$ — Move the decimal points to make the divisor a whole number.

$12.\overline{)08.4}$ — Divide. Add a decimal point to the quotient.

Solve each problem. Use mental math.

1. $0.036 \div 0.6 =$
2. $0.55 \div 0.005 =$
3. $7.2 \div 1.2 =$
4. $100 \div 0.01 =$
5. $4.8 \div 0.06 =$
6. $0.0027 \div 0.9 =$
7. $1.69 \div 0.13 =$
8. $0.108 \div 0.09 =$
9. $0.44 \div 0.4 =$
10. $1.21 \div .11 =$
11. $8.4 \div 0.12 =$
12. $0.064 \div 0.8 =$
13. $3.6 \div 0.009 =$
14. $0.0054 \div 0.006 =$
15. $0.012 \div 0.3 =$
16. $14.4 \div 1.2 =$
17. $0.56 \div 0.008 =$
18. $2.6 \div 0.02 =$

Name ____________________ 5.NBT.B.7, 6.NS.B.3

Dividing Decimals

Remember:
To divide decimals, move the decimal points to make the divisor a whole number.
Then, divide. Add a decimal to the quotient.

Solve each problem. Be sure to mark repeating digits by placing a line above the digits that repeat.

1. 2.34 ÷ 1.6 =

2. 2.62 ÷ 0.54 =

3. 23.65 ÷ 22.5 =

4. 872.6 ÷ 2.4 =

5. 1.32 ÷ 1.8 =

6. 3.6 ÷ 0.3 =

7. 44.34 ÷ 32.76 =

8. 34.96 ÷ 3.549 =

9. 7.569 ÷ 3.459 =

10. 8.37 ÷ 4.50 =

11. 9.2 ÷ 1.2 =

12. 16 ÷ 0.48 =

13. 0.014 ÷ 0.005 =

14. 10.81 ÷ 9.1 =

15. 16.25 ÷ 0.5 =

16. 0.564 ÷ 1.2 =

17. 8.704 ÷ 3.4 =

18. 4.848 ÷ 0.08 =

19. 0.0448 ÷ 0.014 =

20. 10.98 ÷ 0.02 =

Name ______________________________ 5.NBT.B.7, 6.NS.B.3

Dividing Decimals

Solve each problem.

1. $6.3056 \div 4.2 =$
2. $7.57 \div 0.1 =$
3. $3.56 \div 2.5 =$
4. $0.493 \div 0.33 =$
5. $8.565 \div 2 =$
6. $0.0135 \div 4.5 =$
7. $40.78 \div 0.2 =$
8. $9.51 \div 3.03 =$
9. $12.63 \div 0.9 =$
10. $9.414 \div 3.3 =$
11. $1.35 \div 0.07 =$
12. $16.73 \div 0.12 =$
13. $3.605 \div 3.2 =$
14. $0.1827 \div 0.09 =$
15. $12.264 \div 5.6 =$
16. $6.65 \div 2.4 =$
17. $2.34 \div 0.012 =$
18. $0.576 \div 4.1 =$
19. $15.8 \div 0.09 =$
20. $8.176 \div 3.2 =$
21. $0.0224 \div 3.6 =$
22. $21.5 \div 0.05 =$

Name ______________________________ (5.NBT.B.7 6.NS.B.3)

Problem Solving with Decimals

John and Jason went to a grocery store and bought some sandwiches for \$4.95, a gallon of fruit juice for \$3.31, and a bag of carrots for \$3.15. How much did they spend altogether?

$$\$4.95 + \$3.31 + \$3.15 = \begin{array}{r} \$4.95 \\ \$3.31 \\ +\ \$3.15 \\ \hline \$11.41 \end{array} \text{ total}$$

Solve each problem. Show your work.

1. George went to the store to buy a pair of pants. The pair of pants that George picked out cost \$45.00. If the price of the pants was reduced by \$10.85, how much will George pay?

2. Gary spent \$13.41 on clothes in January, \$25.95 in February, and \$31.50 in March. Altogether, how much money did he spend on clothes in these months?

3. Jesse buys a shirt for \$32.95 and a pair of shoes for \$46.25. How much money does Jesse spend in all?

4. Heather bought some new fishing equipment. She bought a tackle box for \$24.95, a fishing pole for \$58.49, a life jacket for \$37.75, and an ice chest for \$57.41. How much money did Heather spend on her equipment?

5. Marisol and Kurt are making sandwiches. The ingredients for the sandwiches cost \$6.07. The sandwiches from the store cost \$8.55. How much money are they saving?

6. Cindy buys a shirt for \$26.70, a pair of jeans for \$47.55, a jacket for \$25.54, and a hat for \$20.11. How much does Cindy spend in all?

Name ______________________________ 5.NBT.B.7, 6.NS.B.3

Problem Solving with Decimals

Solve each problem. Show your work.

1. Ellen wanted to buy the following items: a DVD player for $49.95, a DVD holder for $19.95, and a personal stereo for $21.95. Does Ellen have enough money to buy all three items if she has $90 with her?

2. John bought a case of soda (24 cans) at the store for $0.31 a can. How much money did John spend at the store?

3. On Juan's road trip, he spent $15.50 for 12.5 gallons of gas. How much money did he pay per gallon?

4. If a car uses 16.4 gallons of gas in 4 hours, how many gallons are used per hour?

5. Rob rode his bike 48.3 miles in 3 hours. How many miles did he bike in 1 hour?

6. Melissa purchased $39.46 in groceries at a store. Melissa paid with a $50 bill. How much change should the cashier give Melissa?

Name ______________________________ (5.NBT.B.7, 6.NS.B.3)

Problem Solving with Decimals

Solve each problem. Show your work.

1. A cylinder is normally 3.45 cm in diameter. If it is remade to be 0.056 cm larger, what is the new size of the cylinder?

2. Norma has $435.82 in her checking account. How much does she have in her account after she makes a deposit of $115.75 and a withdrawal of $175.90?

3. Alan is connecting three garden hoses to make one longer hose. One hose is 6.25 feet long, another hose is 5.755 feet long, and the last hose is 6.5 feet long. How long is the hose after they are all connected?

4. Amanda lives on a farm out in the country. It takes her 2.25 hours to drive to town and back. She usually goes to town twice a week to get supplies. How much time does Amanda spend driving if she makes 8 trips to town each month?

5. Destiny's favorite apples are $2.50 per pound at the grocery store. She bought 3.5 pounds of the apples. How much did she spend?

6. Jarvis bought a van that holds 21.75 gallons of gas and gets an average of 12.5 miles per gallon. How many miles can he expect to travel on a full tank?

7. Virginia has 12.25 cups of cupcake batter. If each cupcake uses 0.75 cups of batter, how many full-sized cupcakes can she make? How much batter will she have left over?

8. Grace worked 38.5 hours last week and earned $325.67. How much did she make per hour? Round your answer to the nearest cent.

Name ______________________________

7.NS.A.2d

Converting Fractions and Decimals

$\frac{1}{5} \longrightarrow 5\overline{)1.00} \longrightarrow \frac{1}{5} = 0.2$
Terminating

$\frac{1}{3} \longrightarrow 3\overline{)1.00} \longrightarrow \frac{1}{3} = 0.\overline{3}$
Repeating

Write each fraction as a decimal. Draw a line above repeating numbers in decimals.

1. $\frac{2}{3}$
2. $\frac{1}{2}$
3. $\frac{4}{33}$
4. $\frac{13}{15}$
5. $\frac{28}{35}$
6. $\frac{6}{15}$

Terminating Decimals

$0.50 = \frac{5}{100} = \frac{1}{2}$

Repeating Decimals

$x = 0.\overline{6}$ ← 1 digit repeats

$x = \frac{6}{9}$

$x = .\overline{23}$ ← 2 digits repeat

$x = \frac{23}{99}$

Write each decimal as a fraction. Write the answer in simplest form.

7. 0.6875
8. $0.\overline{35}$
9. 0.212
10. $0.\overline{48}$
11. 0.625
12. 0.54

Name ______________________________ 7.NS.A.2d

Converting Fractions and Decimals

Write each fraction as a decimal. Draw a line above repeating numbers in decimals.

1. $\frac{9}{36}$

2. $\frac{8}{15}$

3. $\frac{30}{45}$

4. $\frac{19}{57}$

5. $\frac{45}{72}$

6. $\frac{21}{36}$

7. $\frac{10}{60}$

8. $\frac{32}{36}$

9. $\frac{56}{63}$

Write each decimal as a fraction. Write the answer in simplest form.

10. 0.345

11. 0.12

12. 0.555

13. 0.300

14. 0.942

15. 0.24

16. $0.34\overline{58}$

17. 0.28

18. $0.13\overline{4}$

Name ________________________________ (7.NS.A.2d)

Converting Fractions and Decimals

Write each fraction as a decimal. Draw a line above repeating numbers in decimals.

1. $\frac{5}{10}$
2. $\frac{12}{18}$
3. $\frac{8}{24}$
4. $\frac{56}{64}$
5. $\frac{48}{74}$
6. $\frac{6}{22}$
7. $\frac{16}{72}$
8. $\frac{35}{55}$
9. $\frac{6}{40}$
10. $\frac{4}{36}$
11. $\frac{13}{39}$
12. $\frac{18}{40}$

Write each decimal as a fraction. Write the answer in simplest form.

13. 0.34
14. 0.342
15. 0.708
16. 0.144
17. 0.65
18. 0.920
19. $0.33\overline{4}$
20. 0.438
21. 0.378
22. 0.82

Name ______________________________ 6.RP.A.1, 6.RP.A.2

Ratios and Rates

A **ratio** is a statement of how two numbers compare. Ratios are used to compare the size of numbers or the rate of something. A **unit rate** is a ratio with a denominator of 1.

Ratios: 3 to 12 → $\frac{3}{12} = \frac{1}{4}$

25:30 → $\frac{25}{30} = \frac{5}{6}$

Rate: $\frac{\text{12 cookies}}{\$6}$ → $\frac{\text{2 cookies}}{\$1}$

Write each ratio as a fraction. Write the answer in simplest form.

1. 66 to 40
2. 130 to 112
3. 110:112
4. 65:35
5. 21 to 84
6. 66:166
7. 30 to 323
8. 197 to 17

Find the unit rate. Round to the nearest hundredth.

9. 294 miles on 10 gallons
10. $0.84 for 12 ounces
11. 4.5 inches of rain in 2 months
12. 400 meters in 24.5 seconds
13. $15 for a half dozen roses
14. $80 for 8 movie tickets
15. 10 limes for $2
16. $2.99 for 6 cans of lemonade

Name ____________________

6.RP.A.1, 6.RP.A.2

Ratios and Rates

A **ratio** shows how two numbers compare. It can use the word *to*, a colon (:), or look like a fraction. A **unit rate** is a ratio with a denominator of 1.

Write each ratio as a fraction. Write the answer in simplest form.

1. 20 to 70
2. 14 to 43
3. 121:108
4. 51:102
5. 34 out of 82
6. 112:224
7. 21 to 45
8. 40: 231

Find the unit rate. Round to the nearest hundredth.

9. 20 people in 4 cars
10. 186 miles in 8 hours
11. 82 hours for 18 projects
12. $950 earned in 5 weeks
13. $2.32 for 16 ounces
14. 578 miles in 9 hours
15. 1200 cars in 5 dealerships
16. 132 miles on 4 gallons

Name ____________________ 6.RP.A.1, 6.RP.A.2

Ratios and Rates

Write each ratio as a fraction. Write the answer in simplest form.

1. $\frac{4}{6}$

2. $\frac{28}{36}$

3. 18 out of 30

4. 14 out of 56

5. 2:10

6. 14:21

7. $\frac{32}{64}$

8. 55:33

9. 25 out of 65

10. 100 out of 170

Find the unit rate. Round to the nearest hundredth.

11. 210.8 miles on 12.4 gallons

12. 2.5 inches of rain in 10 hours

13. $10.20 for 15 pounds

14. $62.50 for 25 tickets

15. $1.50 for 25 sticks of gum

16. $3.00 for 5 muffins

17. $4.49 for 16 ounces

18. 240 miles on 8 gallons

19. $425 for 5 days of work

20. 90 miles in 2 hours

Name ______________________________ (7.RP.A.2a, 7.RP.A.2c)

Proportions

A **proportion** is an equation that names 2 equivalent ratios.
Use cross-products to solve proportions.

$$\frac{n}{3} = \frac{6}{9}$$

$$n \times 9 = 3 \times 6$$

$$\frac{9n}{9} = \frac{18}{9}$$

$$n = 2$$

Solve each proportion. Use cross-products.

1. $\frac{3}{2} = \frac{n}{42}$

2. $\frac{2}{5} = \frac{14}{x}$

3. $\frac{n}{16} = \frac{5}{8}$

4. $\frac{7}{5} = \frac{n}{20}$

5. $\frac{3}{4} = \frac{x}{16}$

6. $\frac{46}{92} = \frac{n}{100}$

7. $\frac{2}{3} = \frac{24}{n}$

8. $\frac{n}{2} = \frac{56}{112}$

9. $\frac{9}{27} = \frac{x}{9.6}$

10. $\frac{25}{5} = \frac{150}{x}$

Name ____________________ 7.RP.A.2a, 7.RP.A.2c

Proportions

$$\frac{2}{6} = \frac{x}{18}$$
$$2 \bullet 18 = 6x$$
$$\frac{36}{6} = \frac{6x}{6}$$
$$6 = x$$

Solve each proportion. Use cross-products.

1. $\frac{1}{4} = \frac{x}{8}$ $= 2$

2. $\frac{20}{30} = \frac{5}{n}$ $= 7.5$

3. $\frac{18}{24} = \frac{12}{x}$ $= 16$

4. $\frac{80}{n} = \frac{48}{20}$ $= 33.333$

5. $\frac{5}{5} = \frac{5x}{5}$ $= 1$

6. $\frac{15}{45} = \frac{3}{x}$ $= 9$

7. $\frac{1.8}{n} = \frac{3.6}{2.8}$ $= 1.4$

8. $\frac{8}{n} = \frac{5}{2}$ $= 3.2$

9. $\frac{8}{6} = \frac{n}{27}$ $= 36$

10. $\frac{144}{6} = \frac{6n}{6}$ $= 24$

11. $\frac{x}{3} = \frac{8}{8}$ $= 3$

12. $\frac{36}{12} = \frac{n}{6}$ $= 18$

13. $\frac{0.14}{0.07} = \frac{n}{1.5}$ $= 3$

14. $\frac{5}{n} = \frac{6}{4}$ $= 3.33\overline{3}$

15. $\frac{4}{5} = \frac{x}{5}$ $= 4$

16. $\frac{16}{48} = \frac{x}{50}$ $= 16.\overline{6}\overline{6}$

Name ______________________________ 7.RP.A.2a, 7.RP.A.2c

Proportions

Solve each proportion. Use cross-products.

1. $\frac{182}{x} = \frac{91}{70}$ = 140

2. $\frac{n}{20} = \frac{73}{10}$ = 146

3. $\frac{186}{84} = \frac{93}{b}$ = 42

4. $\frac{84}{136} = \frac{21}{a}$ = 34

5. $\frac{c}{190} = \frac{29}{95}$ = 58

6. $\frac{93}{x} = \frac{31}{65}$ = 195

7. $\frac{a}{134} = \frac{30}{67}$ = 60

8. $\frac{122}{a} = \frac{61}{100}$ = 200

9. $\frac{x}{124} = \frac{92}{62}$ = 184

10. $\frac{80}{188} = \frac{c}{94}$ = 40

11. $\frac{12}{55} = \frac{b}{165}$ = 36

12. $\frac{c}{46} = \frac{164}{92}$ = 82

13. $\frac{b}{1} = \frac{15}{5}$ = 3

14. $\frac{87}{29} = \frac{174}{x}$ = 58

15. $\frac{94}{18} = \frac{x}{36}$ = 188

16. $\frac{d}{35} = \frac{3}{105}$ = 1

17. $\frac{86}{73} = \frac{x}{146}$ = 172

18. $\frac{97}{95} = \frac{a}{190}$ = 194

19. $\frac{99}{43} = \frac{198}{d}$ = 86

20. $\frac{65}{b} = \frac{195}{111}$ = 37

Name ______________________________ 6.RP.A.3, 7.RP.A.1

Problem Solving with Proportions

To solve problems with proportions, be sure to have the same unit in both numerators and the same unit in both denominators.

If a 9-pound turkey takes 180 minutes to cook, how long would a 6-pound turkey take to cook?

$$\frac{\text{pounds}}{\text{time}} = \frac{9}{180} = \frac{6}{m}$$

$$9m = 180 \bullet 6$$

$$\frac{9m}{9} = \frac{1080}{9} \qquad m = 120 \text{ minutes}$$

Solve each problem. Round each answer to the nearest cent or hundredth.

1. There are 220 calories in 4 ounces of beef. How many calories are there in 5 ounces?

   275

2. If John can buy 8 liters of soft drinks at the store for $6.40, how much does it cost him to buy 12 liters?

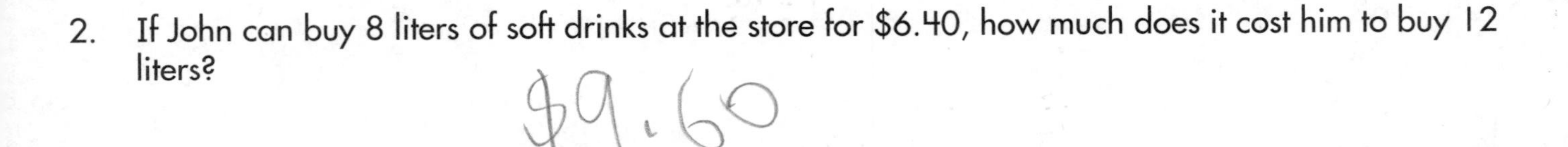

3. Sherri bought a pack of pens that contained 15 pens. How many packs should she buy if she needs 240 pens?

   16

4. Steve won his election by a margin of 7 to 2. His opponent had 3,492 votes. How many votes did Steve have?

   12,222

5. A car traveled 325 miles in 5 hours. How far did the car travel in 9 hours?

   585 hrs

6. A recipe asks for $1\frac{1}{2}$ cups of chocolate chips for 60 cookies. How many cups would be needed for 36 cookies?

   0.9 cups

Name ______________________________

Problem Solving with Proportions

If 3 liters of juice cost $3.75, how much does 9 liters cost?

$$\frac{\text{liters}}{\text{cost}} = \frac{3}{3.75} = \frac{9}{x}$$

$$3x = 3.75 \cdot 9$$

$$\frac{3x}{3} = \frac{33.75}{3}$$

$x = 11.25$
9 liters cost $11.25

Solve each problem. Round each answer to the nearest hundredth.

1. If 3 square feet of fabric cost $3.75, what would 7 square feet cost?

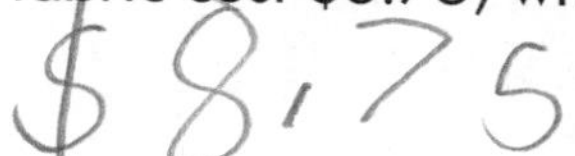

2. A 12-ounce bottle of soap costs $2.50. How many ounces would be in a bottle that costs $3.75?

3. Four pounds of apples cost $5.00. How much would 10 pounds of apples cost?

4. A 12-ounce can of lemonade costs $1.32. How much would a 16-ounce can of lemonade cost?

5. J & S Jewelry company bought 800 bracelets for $450.00. How much did each bracelet cost?

6. A dozen peaches costs $3.60. How much did each peach cost?

7. A 32-pound box of cantaloupe costs $24.40. How much would a 12-pound box cost?

8. If a 10-pound turkey costs $20.42, how much would a 21-pound turkey cost?

Name ______________________________ 6.RP.A.3, 7.RP.A.1

Problem Solving with Proportions

Solve each problem. Round each answer to the nearest hundredth.

1. Buying 19 pounds of apples costs $209. How much would 49 pounds cost?

2. A car can travel 273 miles on 7.8 gallons of gasoline. How far can it travel on 10.4 gallons?

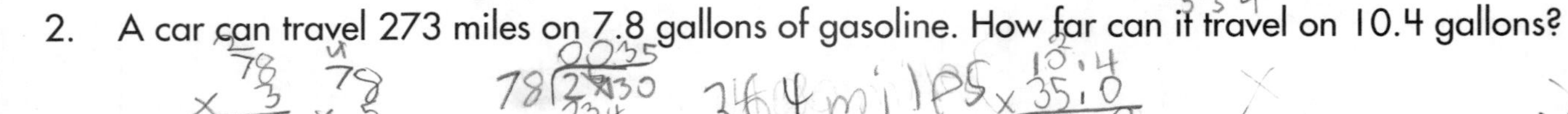

3. A boat can travel 83.82 miles on 27.94 gallons of gasoline. How much gasoline will it need to go 398.1 miles?

4. A boat travels 381 miles in 27.2 hours. How far can it travel in 11.4 hours?

5. Buying 8 pounds of apples costs $48.72. How much would 20 pounds cost?

6. Buying 36 pounds of bananas costs $136.80. How much would 1 pound cost?

7. A boat travels 319 miles in 64 hours. How far can it travel in 77 hours?

8. A boat travels 432.08 miles in 31 hours. How much time will it take for the boat to travel 480.26 miles?

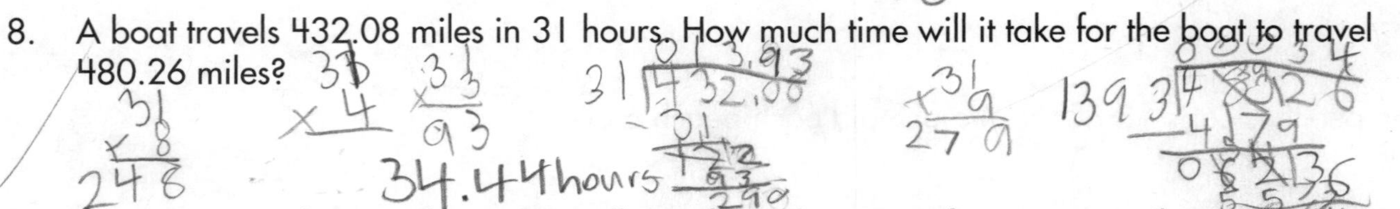

9. A car can travel 210 miles on 10 gallons of gasoline. How far can it travel on 18 gallons?

10. A boat travels 458.4 miles in 13.9 hours. How much time will it take for the boat to travel 493.6 miles?

Name ______________________

7.NS.A.2d

Percents

A percent is a ratio that compares a number to 100.

Fraction to Percent

$$\frac{1}{2} = \frac{x}{100}$$

$$2x = 100$$

$$x = 50$$

$$\frac{1}{2} = 50\%$$

Percent to Fraction

$$12\% = \frac{12}{100}$$

$$\frac{12}{100} = \frac{3}{25}$$

Decimal to Percent

$0.425 = 42.5\%$

When converting a decimal to a percent, multiply by 100, or move the decimal two places to the right.

Percent to Decimal

$68\% = 0.68$

When converting a percent to a decimal, divide by 100, or move the decimal two places to the left.

Write each fraction or decimal as a percent. Round each answer to the nearest hundredth.

1. $\frac{7}{21}$

2. 10.8

3. 12.392

4. 523.32

Write each percent as a decimal and a fraction. Write fractions in simplest form.

5. 40%

6. 30.5%

7. 73%

8. 8.66%

Name ________________________

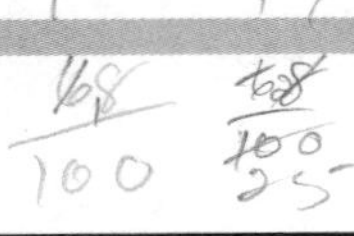

Percents

$$60\% = \frac{60}{100} = \frac{6}{10} = \frac{3}{5}$$

$$51.5\% = \frac{51.5}{100} = \frac{515}{1,000} = \frac{103}{200}$$

Write each percent as a fraction in simplest form. Write each fraction as a percent rounded to the nearest hundredth.

1. 0.68%
2. 33.8%
3. 8.6%
4. 21.98%
5. 4.934%
6. $7\frac{4}{21}$
7. 5.75%
8. $3\frac{34}{88}$
9. $4\frac{5}{46}$
10. $2\frac{5}{7}$
11. 364.69%
12. $21\frac{7}{32}$
13. 12.4%
14. $6\frac{1}{2}$
15. $1\frac{3}{4}$
16. 2.98%

Name ________________________________ 7.NS.A.2d

Percents

Write each fraction or decimal as a percent rounded to the nearest hundredth.

1. 2.3839

2. $\frac{12}{19}$

3. $\frac{5}{46}$

4. $\frac{11}{23}$

5. $\frac{4}{13}$

6. 2.32

7. 17.45

8. 5.293

9. 0.625

10. $\frac{3}{2}$

Complete the table by naming the fraction and decimal for each percent. Write fractions in simplest form.

	Percent	Fraction	Decimal
11.	25%		
12.	5%		
13.	28.5%		
14.	8.75%		
15.	150%		
16.	60%		
17.	62.5%		
18.	48%		
19.	0.3%		
20.	0.6%		

Name ______________________ 7.RP.A.3

Problem Solving with Percents

50% of 60 = ______	______% of 40 = 20	40% of ______ = 30
$\frac{50}{100} = \frac{x}{60}$	$\frac{x}{100} = \frac{20}{40}$	$\frac{40}{100} = \frac{30}{x}$
$100x = 3000$	$40x = 2000$	$40x = 3000$
$x = 30$	$x = 50$	$x = 75$

Solve each problem using proportions.

1. 20% of 15 = ______

2. 30% of 60 = ______

3. 15% of 75 = ______

4. 37% of 65 = ______

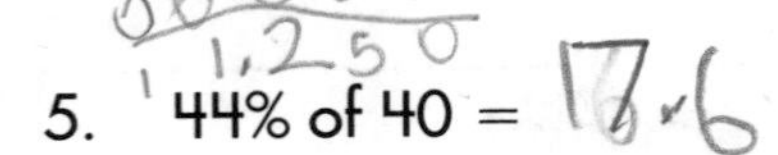

5. 44% of 40 = ______

6. ______% of 35 = 15

7. ______% of 70 = 20

8. ______% of 48 = 8

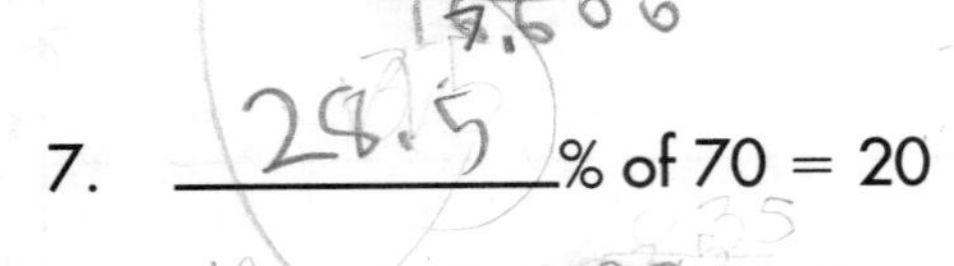

9. 50% of ______ = 32.5

10. 45% of ______ = 4.05

Name ______________________________ 7.RP.A.3

Problem Solving with Percents

To solve word problems with percents, decide what you need to find the percent of by reading the problem. Then, use proportions to solve.

Each item at a sale was reduced by 25%. What was the regular price of a shirt that is reduced $9?

25% of regular price = $9

$$\frac{25}{100} = \frac{9}{r}$$

$$\frac{25r}{25} = \frac{900}{25}$$

$$r = \$36$$

Solve each problem using proportions.

1. 12% of 633.333 = 76
2. 65% of 123.08 = 80
3. 20% of 375 = 75
4. 45% of 333.333 = 150
5. 22% of 154.5454 = 34
6. 50% of 104 = 52
7. Molly made $30 in tips from her customers. If the total of her customers' bills was $200, what percent did her customers tip?

    15%

8. How many problems did Robert get right out of 40 if he received an 87.5% on his test?

    35 questions

9. Mary has sold 90 boxes of cookies. If her goal was to sell 120 boxes, what percentage of her goal has she sold?

    75%

10. How much did Tom pay in income tax on a gross income of $50,000 if 9% of his income was taxed?

    $4500

Name ______________________________ (7.RP.A.3)

Problem Solving with Percents

Solve each problem using proportions.

1. ___50___% of 90 = 45

2. ___19___% of 100 = 19

3. 75% of 60 = ___45___

4. 35% of ___20___ = 7

5. How much is a $48 shirt that is on sale for 25% off?

6. Janet borrowed $5,500 from the bank at an interest rate of 7.5% for 1 year. Assuming she pays it back on time, how much interest will she pay?

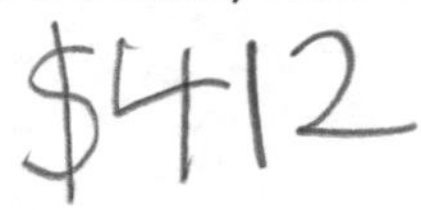

7. If 14 out of 24 students in a class are boys, what percent of the class are girls?

8. At Pinetop Middle School, 372 students ride the bus to school. If this number is 60% of school enrollment, then how many students are enrolled at the school?

9. If you purchase a game console that costs $625, how much sales tax will you pay if the rate is 7.375%?

$46.09

10. A volleyball team won 75% of 80 games played. How many games did they win?

Name ____________________

6.NS.C.5, 6.NS.C.6, 6.NS.C.7

Integers and Absolute Value

Integers are positive and negative values. On a number line, positive integers are whole numbers to the right of 0, and negative integers are whole numbers to the left of 0.

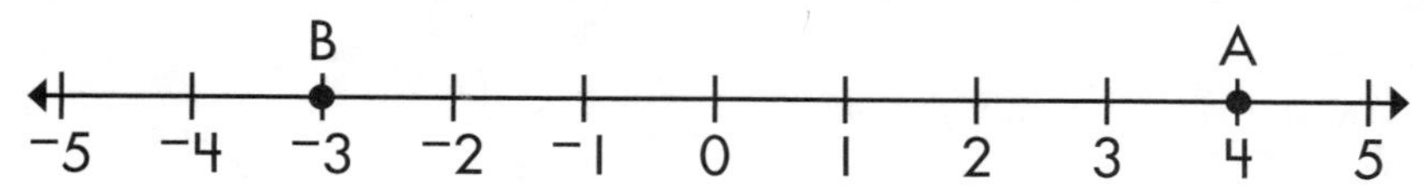

The coordinate of A on this number line is 4 and the coordinate of B on this number line is ⁻3.

The **absolute value** of a number is the distance the number is from 0. It is shown by putting vertical lines on either side of an integer like this: |⁻7| The absolute value of this integer is 7 since it is 7 spaces from 0. Absolute value is always represented by a positive number.

Name the coordinate of each point on the number line using integers.

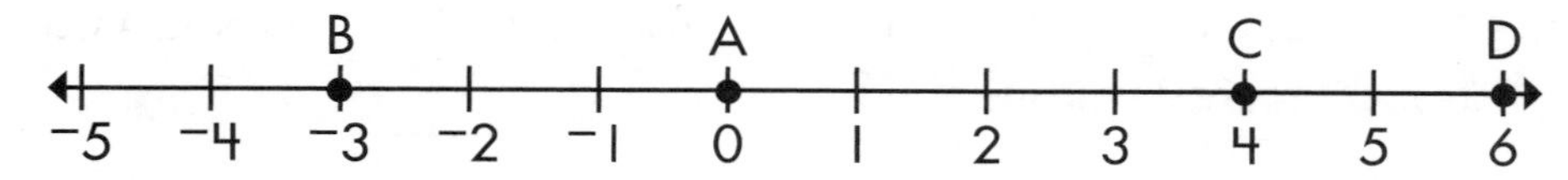

1. A ______
2. B ______
3. C ______
4. D ______

Graph each integer on the number line below.

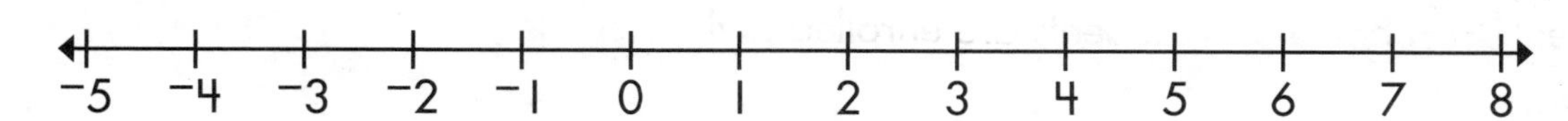

5. A = ⁻4
6. B = 2
7. C = 5
8. D = 7

Write the absolute value of each number.

9. |⁻12|
10. |0|
11. |−10|
12. |⁻15|

Name ______________________________ 6.NS.C.5, 6.NS.C.6, 6.NS.C.7

Integers and Absolute Value

Remember, positive integers are whole numbers to the right of 0 (point B), and negative integers are whole numbers to the left of 0 (point A).

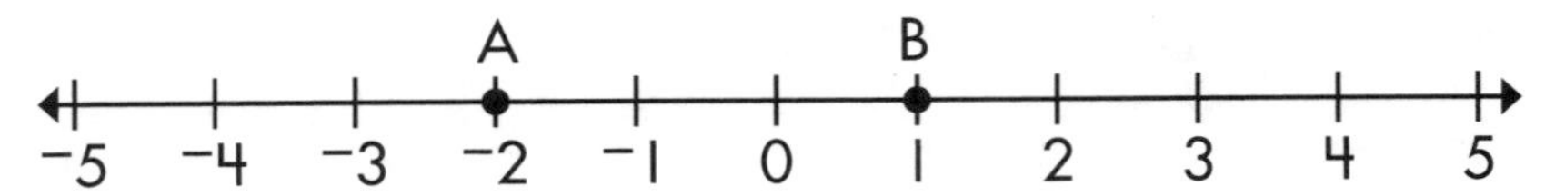

When an integer is shown with 2 negative signs, like $^{-}(^{-}2)$, it represents a positive value.

If absolute value is shown with a negative sign outside of the absolute value, $-|4|$, then the absolute value is $^{-}4$.

Name the coordinate of each point on the number line using integers.

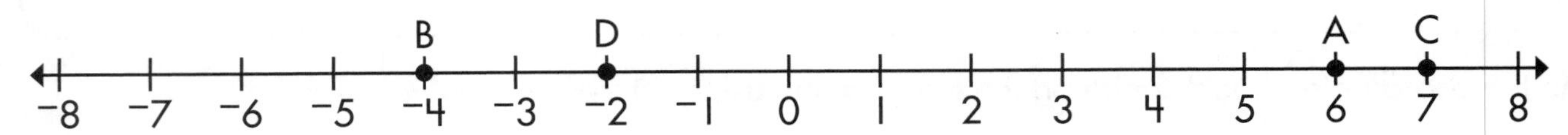

1. A ______

2. B ______

3. C ______

4. D ______

Graph each integer on the number line below.

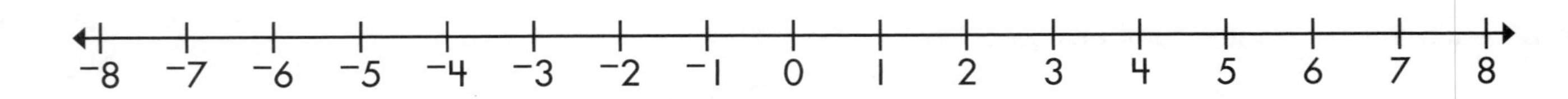

5. $A = ^{-}(^{-}3)$

6. $B = 7$

7. $C = ^{-}1$

8. $D = ^{-}(3)$

Write the absolute value of each number.

9. $-|^{-}15|$

10. $|5|$

11. $|^{-}23|$

12. $|12|$

Name ______________________________ (6.NS.C.5, 6.NS.C.6, 6.NS.C.7)

Integers and Absolute Value

Name the coordinate of each point on the number line using integers.

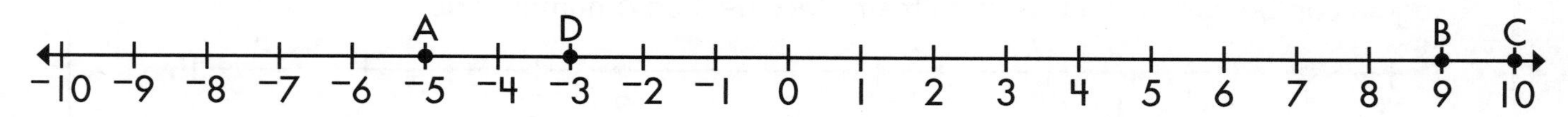

1. A ______

2. B ______

3. C ______

4. D ______

Graph each integer on the number line below.

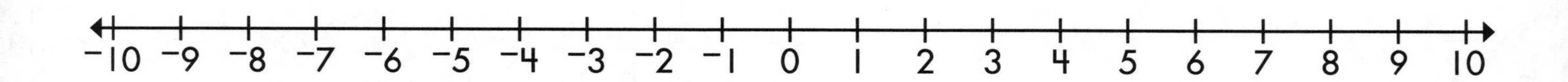

5. $A = -(-5)$

6. $B = 8$

7. $C = -4$

8. $D = -(7)$

9. $E = -(2)$

10. $F = -(-4)$

Write the absolute value of each number.

11. $-|38|$

12. $-|-132|$

13. $|-96|$

14. $|45|$

15. $-|-73|$

16. $-|18|$

Name ____________________ 6.NS.C.7a

Comparing and Ordering Integers

Integers can be compared based on their placement on a number line.

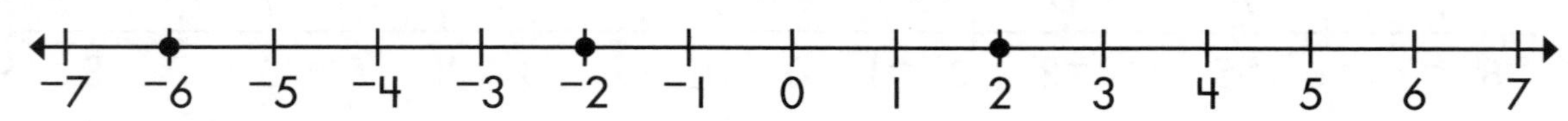

$^{-}6 < 2$ $\quad$ $^{-}6$ is less than 2 because it is to the left on the number line.

$2 > {}^{-}6$ $\quad$ 2 is greater than $^{-}6$ because it is to the right on the number line.

Numbers can also be ordered based on where they fall on a number line.

Use $<$, $>$, or $=$ to compare the numbers.

1. $10 \bigcirc {}^{-}12$

2. $|{}^{-}3| \bigcirc |3|$

3. $^{-}11 \bigcirc {}^{-}2$

4. $0 \bigcirc |{}^{-}2|$

5. $^{-}5 \bigcirc 0$

6. $^{-}8 \bigcirc {}^{-}14$

Order the numbers from least to greatest.

7. $1, {}^{-}4, {}^{-}2$

8. $76, {}^{-}72, 45$

9. $3, {}^{-}3, 4$

10. $^{-}6, 6, {}^{-}7$

Name ______________________________ 6.NS.C.7a

Comparing and Ordering Integers

To compare or order integers, consider their placement on a number line.

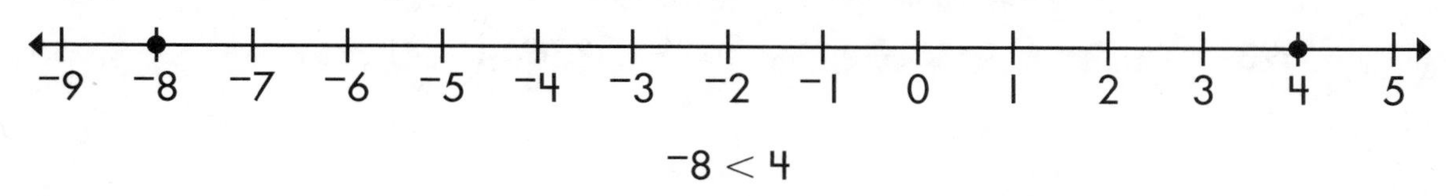

$^{-}8 < 4$

Use $<$, $>$, or $=$ to compare the numbers.

1. $^{-}76$ ○ 82
2. 23 ○ $^{-}13$
3. 35 ○ 15
4. $^{-}66$ ○ $^{-}42$
5. $^{-}81$ ○ 73
6. 615 ○ 413
7. 93 ○ $^{-}52$
8. $^{-}57$ ○ $^{-}32$

Order the numbers from least to greatest.

9. $^{-}82$, $^{-}20$, 73, 29
10. $^{-}49$, 11, 55, 12
11. 16, 67, $^{-}51$, $^{-}3$
12. $^{-}58$, $^{-}14$, $^{-}63$, 3
13. $^{-}77$, 90, $^{-}22$, 8
14. 42, 59, 73, 48
15. $^{-}20$, $^{-}17$, $^{-}15$, $^{-}43$
16. $^{-}94$, 70, $^{-}18$, $^{-}13$

Name ______________________________ 6.NS.C.7a

Comparing and Ordering Integers

Use $<$, $>$, or $=$ to compare the numbers.

1. −791 ◯ 493
2. 836 ◯ 259
3. −71 ◯ −28
4. 916 ◯ −973
5. −33 ◯ −15
6. 286 ◯ −705
7. −973 ◯ −751
8. 56 ◯ −85
9. −254 ◯ 450
10. 619 ◯ 517

Order the numbers from least to greatest.

11. 67, −85, −52, −63
12. −5, 65, −57, 64
13. −2, 48, −31, −88
14. 38, −29, −98, 77
15. 29, 8, −56, 55
16. −75, 63, 6, 50
17. −87, −58, −9, 59
18. −51, 37, −57, −67
19. −64, −87, 87, −85
20. −60, −14, 9, −41

Name ____________________ 7.NS.A.1

Adding Integers

To add integers that have the same sign, add their absolute values. Give the sum the same sign as the integers.

$3 + 5$	$^{-}4 + (^{-}7)$
$\lvert 3\rvert + \lvert 5\rvert$	$\lvert ^{-}4\rvert + \lvert ^{-}7\rvert$
8	$^{-}11$

To add integers that have different signs, subtract their absolute values. Give the result the same sign as the integer with the greater absolute value.

$^{-}7 + 5$
$\lvert ^{-}7\rvert + \lvert 5\rvert$
$7 - 5$
$^{-}2$

Find each sum.

1. $^{-}3 + 5 =$
2. $16 + (^{-}16) =$
3. $25 + 45 =$
4. $^{-}150 + 125 =$
5. $4 + (^{-}11) =$
6. $^{-}11 + (^{-}12) =$
7. $17 + (^{-}6) =$
8. $^{-}8 + 3 =$
9. $9 + (^{-}4) =$
10. $^{-}13 + (^{-}10) =$
11. $^{-}12 + (^{-}8) =$
12. $^{-}5 + (^{-}15) =$

Name ____________________

7.NS.A.1

Adding Integers

$5 + 5$ $\lvert 5 \rvert + \lvert 5 \rvert$ 10	$^-3 + (^-10)$ $\lvert ^-3 \rvert + \lvert ^-10 \rvert$ $^-13$
$5 + (^-13)$ $\lvert 5 \rvert + \lvert ^-13 \rvert$ $13 - 5$ $^-8$	$^-12 + 14$ $\lvert ^-12 \rvert + \lvert 14 \rvert$ $14 - 12$ 2

Find each sum.

1. $8 + (^-11) =$
2. $(^-12) + (^-7) =$
3. $(^-14) + (^-9) =$
4. $^-43 + 68 =$
5. $^-17 + 6 =$
6. $142 + 374 =$
7. $^-21 + (^-34) =$
8. $24 + (^-67) =$
9. $^-73 + 12 =$
10. $5 + 6 =$
11. $88 + (^-34) =$
12. $(^-23) + (^-48) =$
13. $^-19 + 39 =$
14. $^-60 + (^-39) =$
15. $29 + 67 =$
16. $^-73 + 25 =$
17. $^-61 + 61 =$
18. $^-42 + (^-36) + (^-22) =$
19. $51 + 87 + 527 =$
20. $^-595 + 630 =$

Name ______________________________ 7.NS.A.1

Adding Integers

Find each sum.

1. $^{-}6{,}607 + 4{,}362 =$

2. $(^{-}21) + (^{-}59) + (^{-}828) =$

3. $23 + 54 + 56 =$

4. $17{,}985 + (^{-}22{,}581) =$

5. $67{,}888 + (^{-}78{,}952) =$

6. $213 + 375 =$

7. $(^{-}16) + (^{-}16) + (^{-}16) =$

8. $48 + (^{-}56) =$

9. $^{-}99 + 94 =$

10. $(^{-}13) + (^{-}65) + (^{-}78) + (^{-}332) =$

11. $12 + 45 + 332 =$

12. $45{,}908 + (^{-}12{,}921) =$

13. $^{-}112{,}956 + 564{,}258 =$

14. $(^{-}20) + (^{-}68) =$

15. $15 + 41 + 7 =$

16. $^{-}34{,}132 + 81{,}323 =$

17. $850 + (^{-}828) =$

18. $^{-}563{,}937 + 76{,}412 =$

19. $(^{-}14) + (^{-}34) + (^{-}67) =$

20. $^{-}34 + (^{-}76) =$

Name ______________________________ 7.NS.A.1

Subtracting Integers

To subtract an integer, add the number's opposite.

$10 - 12$	Change to addition and add the opposite of 12.
$10 + (^-12)$	
$\lvert ^-12 \rvert + \lvert 10 \rvert$	
$12 - 10$	Subtract the absolute values.
$^-2$	The answer has a negative sign since 12 is greater than 10.

Solve each problem.

1. $1 - 3 =$

2. $10 - (^-5) =$

3. $^-8 - 14 =$

4. $^-4 - 7 =$

5. $^-2 - (^-12) =$

6. $20 - 18 =$

7. $^-10 - 6 =$

8. $12 - (^-9) =$

9. $^-13 - (^-8) =$

10. $7 - 16 =$

11. $24 - (^-26) =$

12. $^-5 - 15 =$

13. $^-40 - 3 =$

14. $23 - 30 =$

Name ______________________________ 7.NS.A.1

Subtracting Integers

$6 - 10 = 6 + (^-10) = ^-4$

Add the opposite

$6 - (^-10) = 6 + 10 = 16$

Add the opposite

Solve each problem.

1. $^-8 - 3 =$
2. $56 - (^-65) =$
3. $^-52 - (^-34) =$
4. $^-19 - (^-13) =$
5. $42 - 23 =$
6. $77 - 22 =$
7. $17 - 26 =$
8. $^-594 - (^-73) =$
9. $^-117 - 29 =$
10. $^-749 - 629 =$
11. $19 - (^-342) =$
12. $2{,}567 - (^-492) =$
13. $5{,}762 - 2{,}144 =$
14. $121 - 154 =$
15. $^-8 - (^-27) =$
16. $^-87 - 129 =$
17. $45 - 75 =$
18. $688 - 456 =$
19. $187 - (^-48) =$
20. $157 - (^-452) =$

Name ______________________ 7.NS.A.1

Subtracting Integers

Solve each problem.

1. 7 − 15 =

2. 319 − (⁻749) =

3. ⁻18 − 6 =

4. ⁻60 − 17 =

5. ⁻45 − (⁻45) =

6. ⁻54 − (⁻95) =

7. ⁻154 − 56 =

8. ⁻625 − 127 =

9. ⁻21 − (⁻45) =

10. 564 − (⁻373) =

11. 3 − (⁻67) =

12. 6,593 − (⁻4,132) =

13. 0 − 15 =

14. 108,762 − (⁻95,671) =

15. ⁻3 − (⁻7) =

16. ⁻9,774 − 8,834 =

17. ⁻5 − (⁻67) =

18. ⁻23 − 56 =

19. ⁻44 − 57 =

20. ⁻475,824 − (⁻153,198) =

21. ⁻36 − 69 =

22. ⁻934 − (⁻672) =

23. ⁻630,805 − (⁻512,156) =

24. 899,342 − (⁻392,231) =

Name ________________________________ 7.NS.A.2a, 7.NS.A.2c

Multiplying Integers

To multiply integers, simply multiply the two numbers. The product of two integers with the same sign is positive and the product of two integers with different signs is negative.

$12 \times 2 = 24$
$^{-}4 \times (^{-}9) = 36$
The product is positive since both integers have the same sign.

$^{-}5 \times 7 = {^{-}35}$ The product is negative since the integers have different signs.

Solve each problem.

1. $7 \times 4 =$

2. $^{-}3 \times (^{-}5) =$

3. $^{-}7 \times 0 =$

4. $6 \times (^{-}8) =$

5. $^{-}9 \times 9 =$

6. $^{-}2 \times (^{-}11) =$

7. $^{-}10 \times 7 =$

8. $^{-}16 \times (^{-}2) =$

9. $^{-}12 \times (^{-}11) =$

10. $^{-}15 \times 0 =$

11. $5 \times 8 =$

12. $8 \times (^{-}10) =$

13. $6 \times (^{-}20) =$

14. $^{-}3 \times 13 =$

Name ______________________________

7.NS.A.2a, 7.NS.A.2c

Multiplying Integers

$\underbrace{(3)(3)}_{\text{}} = 9$ $\underbrace{(^-2)(^-3)} = 6$ — like signs = positive

$\underbrace{(^-3)(3)} = {}^-9$ $\underbrace{(^-2)(3)} = {}^-6$ — unlike signs = negative

When multiplying more than 2 integers, an even number of negative signs will give a positive answer. An odd number of negative signs will give a negative answer.

Solve each problem.

1. $(33)(^-123)(12) =$
2. $(^-434)(^-7) =$
3. $(15)(^-4) =$
4. $(^-5)(^-28)(^-23) =$
5. $(30)(5) =$
6. $(13)(^-28) =$
7. $(^-72)(43) =$
8. $(^-3)(9) =$
9. $(56)(12) =$
10. $(14)(^-33)(2) =$
11. $(32)(^-48) =$
12. $(20)(^-3)(23)(^-3) =$
13. $(^-39)(^-58) =$
14. $(12)(^-12)(2)(^-33) =$
15. $(^-20)(^-10)(2)(3) =$
16. $(37)(^-90) =$
17. $(121)(^-10)(21) =$
18. $(^-9)(^-88)(^-7) =$
19. $(^-13)(^-13) =$
20. $(^-32)(^-22)(^-45) =$

Name ______________________________ 7.NS.A.2a, 7.NS.A.2c

Multiplying Integers

Solve each problem.

1. $(8)(^{-}99)(^{-}22)(^{-}7) =$

2. $(^{-}9)(^{-}82)(^{-}7) =$

3. $(^{-}7)(3) =$

4. $(^{-}6)(^{-}9) =$

5. $(^{-}10)(^{-}5)(^{-}3) =$

6. $(5)(^{-}4)(^{-}3) =$

7. $(^{-}7)(^{-}2)(^{-}5) =$

8. $(^{-}7)(^{-}14)(144) =$

9. $(^{-}17)(^{-}2) =$

10. $(^{-}2)(^{-}13)(^{-}4) =$

11. $(^{-}8)(^{-}9) =$

12. $(^{-}1)(22)(^{-}33)(44) =$

13. $(21)(^{-}22) =$

14. $(^{-}85)(^{-}215) =$

15. $(4)(111)(^{-}1) =$

16. $(213)(4)(18) =$

17. $(^{-}19)(^{-}38) =$

18. $(^{-}5)(^{-}100)(^{-}302) =$

19. $(1)(^{-}41)(^{-}6) =$

20. $(^{-}33)(213) =$

Name ________________________________

7.NS.A.2b, 7.NS.A.2c

Dividing Integers

To divide integers, simply divide the two numbers. The quotient of two integers with the same sign is positive and the quotient of two integers with different signs is negative.

$^-10 \div (^-5) = 2$
$60 \div 5 = 12$
} The quotient is positive since both integers have the same sign.

$^-45 \div 9 = {}^-5$ The quotient is negative since the integers have different signs.

Solve each problem.

1. $9 \div 3 =$

2. $45 \div (^-15) =$

3. $^-28 \div 4 =$

4. $0 \div (^-7) =$

5. $^-16 \div (^-8) =$

6. $^-26 \div (^-13) =$

7. $^-121 \div (^-11) =$

8. $0 \div 15 =$

9. $100 \div (^-20) =$

10. $20 \div (^-5) =$

11. $(^-9) \div 3 =$

12. $^-38 \div (^-19) =$

13. $96 \div 16 =$

14. $^-21 \div (^-7) =$

Name ______________________________

7.NS.A.2b, 7.NS.A.2c

Dividing Integers

$\frac{-18}{-3} = 6$
like signs = positive

$24 \div (-4) = -6$
unlike signs = negative

Solve each problem.

1. $100 \div (-4) =$

2. $\frac{-18}{18} =$

3. $-60 \div 3 =$

4. $\frac{-104}{8} =$

5. $120 \div (-6) =$

6. $\frac{-77}{7} =$

7. $88 \div (-22) =$

8. $\frac{36}{-9} =$

9. $-188 \div 4 =$

10. $\frac{168}{21} =$

11. $144 \div (-12) =$

12. $\frac{-50}{-5} =$

13. $80 \div (-5) =$

14. $-36 \div 6 =$

15. $72 \div 4 =$

16. $\frac{169}{-13} =$

17. $\frac{210}{-10} =$

18. $\frac{-20}{4} =$

19. $-150 \div 6 =$

20. $\frac{-288}{-12} =$

Name ______________________________ 7.NS.A.2b, 7.NS.A.2c

Dividing Integers

Solve each problem.

1. $^{-}14 \div 14 =$

2. $\frac{-77}{11} =$

3. $60 \div (^{-}10) =$

4. $^{-}160 \div (^{-}40) =$

5. $^{-}72 \div 9 =$

6. $\frac{-80}{10} =$

7. $\frac{-755}{-5} =$

8. $\frac{-72}{8} =$

9. $^{-}54 \div (^{-}9) =$

10. $\frac{-35}{-7} =$

11. $^{-}195 \div (^{-}65) =$

12. $\frac{-468}{26} =$

13. $^{-}150 \div (^{-}50) =$

14. $\frac{-253}{11} =$

15. $189 \div (^{-}21) =$

16. $\frac{66}{-2} =$

17. $75 \div (^{-}3) =$

18. $\frac{-84}{-7} =$

19. $^{-}210 \div (^{-}5) =$

20. $\frac{-552}{-23} =$

21. $^{-}94 \div 2 =$

22. $\frac{-310}{5} =$

23. $^{-}125 \div 5 =$

24. $\frac{-258}{-3} =$

Name ________________________________ 7.EE.A.1

Properties of Operations

The **commutative property** states that the order of the operation can be changed and the result is still the same. Addition and multiplication both have the commutative property.

$2 + 3 = 3 + 2$ $4 \times 2 = 2 \times 4$

The **associative property** states that a set of numbers can be grouped in different ways without changing the result. Addition and multiplication both have the associative property.

$(3 + 5) + 7 = 3 + (5 + 7)$ $(12 \times 13) \times 14 = 12 \times (13 \times 14)$

The **distributive property** allows for multiples before operations to be distributed and for factors within multiples to be taken out.

$3(7 + 2) = (3)(7) + (3)(2)$
$21 + 6 = 27$

$3x + x = 24$
$x(3 + 1) = 24$

Rewrite each expression using the commutative property.

1. $6 + y = 12$

2. $a + 17 = 21$

3. $x + 4 = 12$

Rewrite each expression using the associative property.

4. $8 + (2 + 17)$

5. $(3 \times {}^{-}4) \times 250$

6. $68 + (32 + 54)$

Rewrite each expression using the distributive property. Do not simplify.

7. $4(12 + 15)$

8. $(10 + 13)z$

9. $7r + 8r + 2$

Name ______________________________ 7.EE.A.1

Properties of Operations

The **commutative property** states that the order of the operation can be changed and the result is still the same.

The **associative property** states that a set of numbers can be grouped in different ways without changing the result.

The **distributive property** allows for multiples before operations to be distributed and for factors within multiples to be taken out.

Rewrite each expression using the commutative property.

1. $h \times 15 = 75$
2. $56 = y \times 7$
3. $11 + k = 32$
4. $3 = r + 8$

Rewrite each expression using the associative property. Then, solve the expression.

5. $(3 \times 25) \times 4$
6. $(21 + 45) + 55$
7. $(12 \times 5) \times 20$
8. $75 + (25 + 19)$

Rewrite each expression using the distributive property.

9. $3a + 6b$
10. $x(6 + 8)$
11. $2(5x + 8y)$
12. $2(b + 4) + 8b$

Name ________________________________ 7.EE.A.1

Properties of Operations

Rewrite each expression using the commutative property.

1. $39 = 13 \times f$

2. $9 + y = 45$

3. $3x + 4y = 18$

4. $8 \times 3b = 456$

5. $9 \times 3 \times 6 = 162$

Rewrite each expression using the associative property. Then, solve the expression.

6. $^{-}20 \times (50 \times {}^{-}29)$

7. $5 \times (10 \times 18)$

8. $(400 + 60) + 98$

9. $(3 \times 20) \times 5$

10. $18 + (9 + 91)$

Rewrite each expression using the distributive property. Simplify when possible.

11. $a(7 + 1) + 15$

12. $k + 5 + 7 + 3k$

13. $2c + 6c + 9(c + 3)$

14. $10(7 \times r)$

15. $(6y + 5y) + 1$

Name ____________________

5.OA.A.1, 6.EE.A.2c

Order of Operations

When an expression contains more than one operation, it is important to use the order of operations to find its value.

1. Solve inside parentheses.
2. Multiply and divide from left to right.
3. Add and subtract from left to right.

$$2(8 + 6) - 7 \times 3$$
$$2(14) - 7 \times 3$$
$$28 - 21$$
$$7$$

Solve each problem using order of operations.

1. $(18 - 2) \div 4 =$

2. $12 - (4 + 7) =$

3. $7 \times (3 + 4) =$

4. $12 \div 3 \times 2 =$

5. $(11 + 4) \div 5 =$

6. $(10 \times 4) \div (2 \times 2) =$

7. $30 \div 6 - 1 =$

8. $42 \div (5 + 2) \times 3 =$

9. $56 \div (7 \times 2) =$

10. $(12 + 8) \div 4 =$

Name ____________________

5.OA.A.1, 6.EE.A.2c

Order of Operations

$$^{-}4 \times 2 + 2 = {}^{-}8 + 2 = {}^{-}6$$

$$2\frac{1}{4} \div (4 + 8) = \frac{9}{8} \div 12 = \frac{9}{8} \times \frac{1}{12} = \frac{9}{96} \text{ or } \frac{3}{32}$$

Solve each problem using order of operations.

1. $2 \times 3\,[7 + (6 \div 2)] =$

2. $\frac{2}{3}(-15 - 4) =$

3. $-8 \div (-2) + 5 \times (\frac{-1}{2}) - 25 \div 5 =$

4. $-30 \div 6 + 4\frac{1}{5} =$

5. $(9\frac{1}{3} + 4\frac{1}{3}) \div 6 - (-12) =$

6. $\frac{[(60 \div 4) + 35]}{(^{-}12 + 35)} =$

7. $\frac{3}{4}\,[(-15 + 4) + (6 + 7) \div (-3)] =$

8. $3[-3(2 - 8) - 6] =$

Name ______________________________ 5.OA.A.1, 6.EE.A.2c

Order of Operations

Solve each problem using order of operations.

1. $3 \times 3[2 - (9 \div 3)] =$

2. $\frac{1}{2}(-12 + 6) =$

3. $(5\frac{1}{5} - 6\frac{1}{5}) \times 6 - (-16) =$

4. $-20 \div 3 + 2\frac{2}{3} =$

5. $-5 \div (-3) - 2 \times (-\frac{1}{3}) - 21 \div 7 =$

6. $\frac{[(20 \div 2) + 10]}{(^{-}10 + 20 + 30)} =$

7. $2[-5(4 - 12) - 3] =$

8. $\frac{1}{2}[(-12 - 2) + (1 + 8) \div (-8)] =$

9. $[(3 \times 3) - (30 \div 6)] + (-27) - 13 =$

10. $2 \div [(4 \div 2) + (-32 \div 8)] =$

11. $30 \times [(3 \times 9) - (21 \div 7)] + (-32) =$

Name ______________________________ 6.EE.A.2a

Using Variables

An expression that contains variables, numbers, and at least one operation is an **algebraic expression**. An algebraic expression can be evaluated by replacing the variables with an assigned value.

If $x = 2$ and $y = 5$:

$$6x - 2y$$
$$6(2) - 2(5)$$
$$12 - 10$$
$$2$$

Evaluate each expression if $x = 5$, $y = 2$, and $z = 8$.

1. $2z - 3x =$

2. $\frac{6x}{y} + z =$

3. $2z - xy =$

4. $10x - (4y + z) =$

5. $4x - (y + z) =$

6. $\frac{7z}{x} + y =$

7. $6z + 7y - 3x =$

8. $2z + 3x + 4y =$

Name ______________________________ 6.EE.A.2a

Using Variables

If $w = \frac{1}{5}$, $x = 4$, and $y = {}^{-}5$,

then $3x(5w + 2y) = 3 \bullet 4[5(\frac{1}{5}) + 2({}^{-}5)] = 12[(1 + ({}^{-}10)] = 12({}^{-}9) = {}^{-}108$

Evaluate each expression if $w = \frac{1}{5}$, $x = 4$, and $y = {}^{-}5$.

1. $y(w + 7) =$

2. $3w + 4(x - y) =$

3. $6[w + ({}^{-}y)] =$

4. $wx + x + 6xy =$

5. $5(w - 2y) =$

6. $w(x + y) =$

7. $w(xw + xy) =$

8. $7w - (xy + 3) =$

9. $3w(3y + 5x) =$

10. $wx(3w + 3y - 6) =$

11. $3w - 4x =$

12. $10y(4y + 2w) =$

13. $8x + ({}^{-}12x) =$

14. $4w - 7x + 3y - 2w =$

Name ______________________________

6.EE.A.2a

Using Variables

Evaluate each expression.

1. $5 + 6k + 2h =$ if $k = {}^-7$ and $h = 9$

2. $\frac{z}{5} - 3f =$ if $z = 20$ and $f = {}^-6$

3. ${}^-2(5r + 6b) =$ if $b = 6$ and $r = 4$

4. ${}^-7(8 - 6z) + 5c =$ if $z = {}^-3$ and $c = 4$

5. $3 + 8({}^-2k - 7z) =$ if $z = {}^-8$ and $k = {}^-2$

6. $9b + k =$ if $b = 7$ and $k = {}^-8$

7. $\frac{{}^-4}{b} + 2x =$ if $b = 2$ and $x = 5$

8. $5 - \frac{16}{s} - 3w =$ if $s = 4$ and $w = {}^-6$

9. $n + 5x =$ if $n = 2$ and $x = {}^-5$

10. ${}^-2(4z - 7b) =$ if $z = 4$ and $b = 3$

11. $6w - 7({}^-9 + 5h) =$ if $h = {}^-4$ and $w = 2$

12. ${}^-9({}^-8z - 2) + 6n =$ if $z = {}^-7$ and $n = 5$

13. ${}^-5(9c + 6s) =$ if $s = {}^-8$ and $c = {}^-7$

14. $\frac{18}{x} - 3 - 8r =$ if $x = 6$ and $r = 2$

15. ${}^-7(6b - 3f) =$ if $b = 4$ and $f = {}^-2$

16. $5s - 8k - 3 =$ if $s = {}^-7$ and $k = 4$

Name ______________________________ 7.EE.B.4a

Solving Addition Equations

To find a solution to an addition equation, the variable must be isolated. Subtract addends from both sides of the equation to isolate the variable. Then, solve.

$$r + 12 = 67$$
$$r + 12 - 12 = 67 - 12$$
$$r = 55$$

Solve each equation for the variable.

1. $x + 7 = 12$

2. $24 = h + 3$

3. $6 + a = 15$

4. $21 + y = 15$

5. $n + 10 = 14$

6. $64 + h = 36$

7. $^{-}13 = z + 7$

8. $4 + r = {}^{-}6$

Name ______________________________

7.EE.B.4a

Solving Addition Equations

$$1.4 = {}^{-}2.4 + x$$
$$1.4 + 2.4 = {}^{-}2.4 + 2.4 + x$$
$$3.8 = 0 + x$$
$$3.8 = x$$

Solve each equation for the variable.

1. $x + ({}^{-}5\frac{3}{4}) = {}^{-}10\frac{1}{4}$

2. ${}^{-}35 = x + 35$

3. $w + 79 = {}^{-}95$

4. $-\frac{1}{4} + x = -\frac{1}{4}$

5. $7 + c = {}^{-}14$

6. ${}^{-}4.5 = 9\frac{1}{2} + c$

7. ${}^{-}21 = y + 18$

8. $22 = c + ({}^{-}13)$

9. ${}^{-}9 + r = 22$

10. $x + ({}^{-}8) = 9$

11. $3.5 = n + 4.6$

12. $-2\frac{1}{2} + k = -3\frac{5}{7}$

13. ${}^{-}2{,}929 + y = 4{,}242$

14. $z + 5.2 = 7.1$

Name ____________________ 7.EE.B.4a

Solving Addition Equations

Solve each equation for the variable.

1. $-\frac{3}{4} + j = ^{-}3\frac{1}{2}$

2. $^{-}8\frac{3}{7} + y = ^{-}9\frac{2}{5}$

3. $^{-}3.1 = 4\frac{3}{4} + e$

4. $b + (^{-}9) = 36$

5. $5.77 = q + 9$

6. $^{-}3{,}282 = n + 1{,}111$

7. $d + 12 = ^{-}18$

8. $f + (^{-}3\frac{1}{4}) = ^{-}7\frac{1}{4}$

9. $x + 9 = ^{-}22$

10. $18 + k = 16$

11. $2 + x = ^{-}20$

12. $(^{-}24.64) + c = ^{-}7.45$

13. $(^{-}4.48) + f = 17.6$

14. $(^{-}22) + r = ^{-}17$

15. $(^{-}1) + g = ^{-}11$

16. $(^{-}17) + p = ^{-}24$

17. $14 + x = 18$

18. $-\frac{5}{2} + w = -\frac{9}{22}$

Name ______________________

7.EE.B.4a

Solving Subtraction Equations

To find a solution to a subtraction equation, the variable must be isolated. Add one value to both sides of the equation to isolate the variable. Then, solve.

$$x - 16 = 32$$
$$x - 16 + 16 = 32 + 16$$
$$x = 48$$

Solve each equation for the variable.

1. $b - 14 = 51$

2. $36 = d - 13$

3. $x - 28 = 72$

4. $11 = x - 1$

5. $^{-}600 = c - (^{-}400)$

6. $y - 8 = 8$

7. $a - 15 = {}^{-}21$

8. $2 - y = {}^{-}4$

Name ____________________

7.EE.B.4a

Solving Subtraction Equations

$$32 = x - (^-8)$$
$$32 = x + 8$$
$$32 - 8 = x + 8 - 8$$
$$24 = x + 0$$
$$24 = x$$

Solve each equation for the variable.

1. $^-2{,}547 = n - 5{,}534$

2. $^-44 = m - 32$

3. $d - (^-8) = 45$

4. $-\frac{1}{3} - g = -\frac{1}{3}$

5. $^-15 = p - 6$

6. $3.65 = n - 7$

7. $34 = b - (^-2)$

8. $a + (^-4\frac{1}{3}) = {}^-15\frac{1}{3}$

9. $f - 16 = {}^-32$

10. $x - 8 = 34$

11. $^-3.4 = h - 8.5$

12. $^-2.2 = 8\frac{4}{5} + d$

13. $-3\frac{2}{3} + k = {}^-6\frac{3}{4}$

14. $z - (^-21.5) = {}^-2.356$

Name ____________________

7.EE.B.4a

Solving Subtraction Equations

Solve each equation for the variable.

1. $11.4 - k = 5.2$

2. $^{-}56 = c - (^{-}8)$

3. $^{-}6.7 = y - 27$

4. $w - (^{-}43.7) = {}^{-}3.674$

5. $31 = d - 12$

6. $y - 1 = \frac{5}{14}$

7. $z - 14 = {}^{-}24$

8. $s - (^{-}3) = {}^{-}22$

9. $v - (^{-}15.69) = 25.6$

10. $k - \frac{15}{13} = \frac{6}{13}$

11. $p - 23 = {}^{-}13$

12. $x - (^{-}4) = {}^{-}19$

13. $e - (^{-}14) = 21$

14. $h - (^{-}13.94) = 21.06$

15. $a - 2 = 10$

16. $j - (^{-}25) = 1$

17. $d - (^{-}11.31) = 6.64$

18. $x - \frac{21}{22} = -\frac{2}{11}$

Name ______________________________ 7.EE.B.4a

Solving Multiplication Equations

To find a solution to a multiplication equation, the variable must be isolated. Divide the known factor from both sides of the equation to isolate the variable. Then, solve.

$$4x = 84$$
$$4x \div 4 = 84 \div 4$$
$$x = 21$$

Solve each equation for the variable.

1. $9x = 63$

2. $^{-}96b = 96$

3. $4d = {}^{-}36$

4. $^{-}64 = {}^{-}16y$

5. $^{-}25x = {}^{-}125$

6. $3x = 21$

7. $6n = {}^{-}72$

8. $12a = 156$

9. $^{-}12r = 12$

10. $54 = {}^{-}9e$

Name ________________________________

7.EE.B.4a

Solving Multiplication Equations

$$4y = {}^{-}24$$
$$4y \div 4 = {}^{-}24 \div 4$$
$$1y = {}^{-}6$$
$$y = {}^{-}6$$

Solve each equation for the variable.

1. $36 = {}^{-}6d$

2. $3m = {}^{-}5$

3. ${}^{-}169 = 13b$

4. $0.24y = 1.2$

5. ${}^{-}15m = 15$

6. $43\frac{1}{2} = {}^{-}13d$

7. ${}^{-}7b = {}^{-}77$

8. ${}^{-}12n = {}^{-}56$

9. $3.5 = 7x$

10. ${}^{-}0.0006 = 0.02c$

11. ${}^{-}2.1 = 0.7c$

12. $33k = {}^{-}878$

13. $1\frac{2}{3} = 9x$

14. $1.44 = 12r$

Name ______________________________ 7.EE.B.4a

Solving Multiplication Equations

Solve each equation for the variable.

1. $^{-}250 = 25s$

2. $18\frac{1}{3} = {}^{-}12w$

3. $72 = {}^{-}8r$

4. $15n = {}^{-}3$

5. $^{-}12q = 12$

6. $^{-}0.0009 = 0.03q$

7. $56 = {}^{-}7e$

8. $43d = {}^{-}734$

9. $^{-}4.5 = 9h$

10. $0.48y = 2.4$

11. $^{-}6.7 = {}^{-}0.137k$

12. $9h = {}^{-}90$

13. $2\frac{4}{6} = 5e$

14. $^{-}13g = {}^{-}78$

Name ________________________________

Solving Division Equations

To find a solution to a division equation, the variable must be isolated. Multiply each side of the equation by the divisor to isolate the variable. Then, solve.

$$f \div 21 = 2$$
$$f \div 21 \times 21 = 2 \times 21$$
$$f = 42$$

Solve each equation for the variable.

1. $\frac{a}{-6} = 2$

2. $\frac{b}{40} = -3$

3. $\frac{r}{15} = 20$

4. $-77 = \frac{d}{11}$

5. $\frac{f}{3.6} = 16$

6. $-6 = \frac{x}{6}$

7. $\frac{h}{9} = 63$

8. $\frac{m}{5} = 22$

9. $\frac{n}{2} = 8$

10. $\frac{b}{12} = -12$

Name ______________________________ 7.EE.B.4a

Solving Division Equations

$$\frac{2}{3}x = 6$$
$$3 \cdot \frac{2}{3}x = 6 \cdot 3$$
$$2x = 18$$
$$x = 9$$

Solve each equation for the variable.

1. $\frac{m}{7} = 42$
2. $^{-}12 = \frac{d}{4}$
3. $0.9 = \frac{k}{81}$
4. $(\frac{1}{7})n = {}^{-}28$
5. $(\frac{3}{4})z = 144$
6. $\frac{r}{17} = {}^{-}23$
7. $\frac{e}{4} = {}^{-}36$
8. $^{-}3 = (\frac{1}{3})x$
9. $^{-}15 = \frac{x}{3}$
10. $\frac{x}{7} = 56$
11. $(\frac{1}{12})c = 0.6$
12. $(\frac{3}{7})h = 4.5$
13. $\frac{x}{4.1} = 18$
14. $(\frac{1}{4})c = {}^{-}8$

Name ____________________ 7.EE.B.4a

Solving Division Equations

Solve each equation for the variable.

1. $\frac{f}{4.44} = -3.63$

2. $\frac{e}{-5} = 5$

3. $\frac{h}{-2} = -10$

4. $\frac{q}{2.45} = 10.67$

5. $\frac{s}{-1.55} = 10.53$

6. $\frac{e}{(\frac{-8}{13})} = \frac{23}{8}$

7. $\frac{k}{5} = -4$

8. $\frac{z}{10.81} = 2.39$

9. $\frac{a}{-2.25} = -6.16$

10. $\frac{e}{(\frac{-1}{25})} = \frac{25}{19}$

11. $\frac{y}{2.86} = 12.22$

12. $\frac{b}{-10} = 2$

13. $\frac{z}{13} = -1$

14. $\frac{m}{-2} = 10$

Name ______________________________ 7.EE.B.4a

Solving Equations with Multiple Operations

To solve a problem with multiple operations, the variable must be isolated. Isolate the variable by using addition, subtraction, multiplication, and division.

$$2y - 10 = 30$$
$$2y - 10 + 10 = 30 + 10$$
$$\frac{2y}{2} = \frac{40}{2}$$
$$y = 20$$

Solve each equation for the variable. Write the answer in simplest form.

1. $14 = 6c - 4$

2. $13n - 13 = {}^{-}12$

3. $5x - 5 = {}^{-}10$

4. $23x - 12 = {}^{-}33$

5. $10 = 3y + 5$

6. ${}^{-}42 = 6b + 8$

7. ${}^{-}23 = 3e - ({}^{-}9)$

8. $16 + 4y = {}^{-}32$

9. ${}^{-}8r - 9 = {}^{-}24$

10. $16 + \frac{r}{2} = {}^{-}11$

Name ______________________

7.EE.B.4a

Solving Equations with Multiple Operations

$$2x + 5 = 3x + 6$$
$$2x - 3x + 5 = 3x - 3x + 6$$
$$^{-}x + 5 = 6$$
$$^{-}x + 5 - 5 = 6 - 5$$
$$^{-}x = 1$$
$$\frac{^{-}x}{^{-}1} = \frac{1}{^{-}1}$$
$$x = {}^{-}1$$

Solve each equation for the variable. Write the answer in simplest form.

1. $4n - 6 = 6n + 14$

2. $^{-}6g + 12 = 2g + 12$

3. $^{-}d + 9 = d + 5$

4. $23b + 9 = 4b + 66$

5. $7y - 7 = 5y + 13$

6. $^{-}8z = 27 + z$

7. $10w + 6 = 6w - 15$

8. $13y - 26 = 7y + 22$

9. $^{-}r - 5 = 1 - 2r$

10. $3m - 8 = 5m + 8$

11. $4x - 7 = 2x + 7$

12. $2e + 20 = 4e - 12$

13. $18 + 2p = 8p - 13$

14. $4h + 10 = 2h - 22$

Name ____________________ 7.EE.B.4a

Solving Equations with Multiple Operations

Solve each equation for the variable. Write the answer in simplest form.

1. $^{-}2j + 6 = 2j - 8$

2. $4h + 6 = 2h - 8$

3. $9w + 9 = 3w - 15$

4. $45 = {}^{-}9(e + 8)$

5. $35 = {}^{-}7d$

6. $12k + 13 = 8k + 33$

7. $6z = 28 - z$

8. $7(9 - 6j) = {}^{-}63$

9. $4(y - 8) = {}^{-}12$

10. $^{-}6(36 - 10b) + 8 = 32$

11. $^{-}14k = 56$

12. $9(8c - 9) = {}^{-}351$

13. $11g = 121$

14. $\frac{m}{2.5} = 22$

15. $24 = 4[(\frac{h}{2}) - 7]$

16. $^{-}6 = \frac{b}{6}$

17. $4g + 12 = 6g - 4$

18. $(\frac{2}{5})\, h = {}^{-}20$

Name ______________________________ 5.OA.A.2, 7.EE.B.3, 7.EE.B.4a

Writing Algebraic Expressions

Use the words in word problems to help you understand which operations should be used.

Three times a number increased by 5	$3x + 5$
A number increased by 3	$x + 3$
A number divided by 2	$x \div 2$ or $\frac{x}{2}$
The product of 2 and 6	2×6

Write the algebraic expression.

1. Two-fifths of a number decreased by 3

2. Twelve times a number decreased by 4

3. Eight times the difference between x and 7

4. The product of 3 and a number increased by 6

5. One-third times a number increased by 7

6. Four increased by 7 times a number

7. Seven times the sum of twice a number and 16

8. Eleven times the sum of a number and 5 times the number

9. Five times a number plus 6 times the number

10. The quotient of a number and 5 decreased by 3

11. Four times the sum of a number and 8

12. A number increased by 9 times the number

Name ______________________________

Writing Algebraic Expressions

Five more than 2 times a number is 21. What is the number?

$$5 + 2x = 21$$
$$5 - 5 + 2x = 21 - 5$$
$$2x = 16$$
$$x = 8$$

Write an equation for each expression and solve.

1. Two times the sum of a number and 5 is 26. What is the number?

2. The quotient of a number and 3 decreased by 6 is 7. What is the number?

3. The product of a number and 4 increased by 7 is 5. What is the number?

4. Three more than 5 times a number is 58. What is the number?

5. Six more than a number is $^{-}31$. What is the number?

6. Two–thirds of a number increased by 3 is 11. What is the number?

7. Ten less than 2 times a number is 24. What is the number?

Name ______________________________

5.OA.A.2, 7.EE.B.3, 7.EE.B.4a

Writing Algebraic Expressions

Write an equation for each expression and solve.

1. One half of a number is 14 more than 2 times the number. Find the number.

2. Forty increased by 4 times a number is 8 less than 6 times the number. Find the number.

3. Nineteen increased by 3 times a number is 4 less than 4 times the number. Find the number.

4. Four times the sum of a number and 3 is 7 times the number decreased by 3. Find the number.

5. Twice a number decreased by 44 is 6 times the sum of the number and 3 times the number. Find the number.

6. Thirty decreased by 3 times a number is 6 less than 3 times the number. Find the number.

7. Twelve increased by 6 times a number is 6 less than 7 times the number. Find the number.

Name ______________________________ 6.EE.A.3

Simplifying Expressions

Simplify expressions by using the distributive property first.

$2(x + 3y) = 2x + 2 \times 3y = 2x + 6y$

If possible, use addition and subtraction to combine like terms.

$4a - 3a + 7z = (4 - 3)a + 7z = a + 7z$

Expand each expression using the distributive property.

1. $4(2r + 6y) =$
2. $2(3p - 3p) =$
3. $^{-}6(2b + 3c) =$
4. $7(^{-}c + 6d) =$
5. $2(x - 12) =$
6. $12(2y + 5w) =$

Combine like terms.

7. $3x + 3y + xy - 6xy + 5x + (^{-}4y) =$
8. $12p + 4pd - 2p + 6pd =$
9. $2x + 3xy + 4x + 5xy + 6x =$
10. $4x - 2x + 6xy + 21x + (^{-}9xy) - 9 =$
11. $^{-}3n + 12 - 4n =$
12. $5e + 6ed + 5d - 7ed + 7 =$

Name ____________________

6.EE.A.3

Simplifying Expressions

Remember, simplify expressions by using the distributive property first. Then, combine like terms when possible.

Expand each expression using the distributive property.

1. $3(2 + r) =$

2. $3(w - 4) =$

3. $8[y + (^-2x)] =$

4. $5(2 + 13y) =$

5. $2k[^-xy + (^-8)] =$

6. $^-7(2x + 9) =$

7. $5(2y + 5x) =$

8. $3(x + 2y + z) =$

Combine like terms.

9. $3xy + 13xy - 12xy =$

10. $2r + 4ry - 5r + 3x - 4ry =$

11. $10ax - 2ax + 12x - 2a + (^-2x) =$

12. $7a + a - 2a + 3ab - ab + 2ab =$

13. $7r + 2r - 4 =$

14. $5m + 2m + 40m + m + 17 =$

15. $23x - 7x + 4x =$

16. $4x + 3y - 3xy + 6x - 2xy =$

Name ______________________________ 6.EE.A.3

Simplifying Expressions

Simplify each expression by using the distributive property and combining like terms.

1. $7(2x + 5y) + 6xy - 6(3xy + 5x) =$

2. $10p + 5pd - 2p + 6pd =$

3. $9x - 4x + 2x + 8(6x + 2x) =$

4. $3(x - 5x) + 2(xy + 7x) + (^-7xy) =$

5. $3c - 4bc + (^-7b) + 3[2bc - (^-b)] =$

6. $^-2a - [^-3(a + 7)] - 4(^-a + b) =$

7. $2t + 12t - 4(n + 4n) =$

8. $6r + 5r - 8p + 6p + 7(2r - 4r) =$

9. $3n(x - y) + 3n(x + y) - 2 =$

10. $3[h - (^-k)] + 2[^-3h + (^-4k)] =$

11. $^-2(g + 5g) + {}^-2[6f - (^-12g)] =$

12. $4(2x + 2y) - 2[3xy - (^-5x)] =$

13. $5m + 3mn - (-9n) + 2(m - n) =$

14. $5xy - 12xy + 12xy - 9(x + y) =$

Name ______________________________ 7.EE.B.4b

Solving Inequalities

$x + 6 < 4$
$x + 6 - 6 < 4 - 6$
$x < -2$

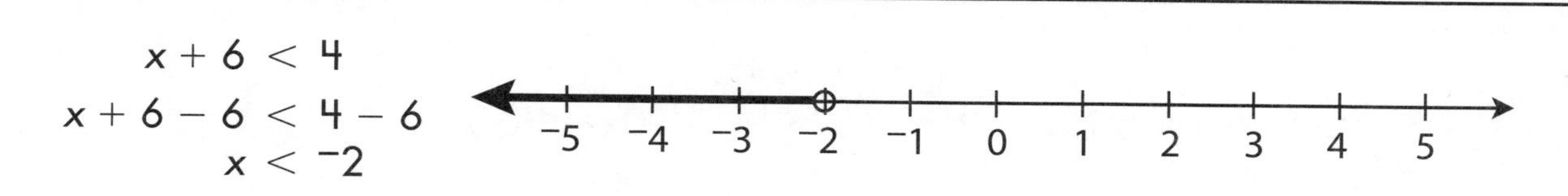

$(-\frac{1}{2})x \geq 2$
$-\frac{2}{1} \bullet (-\frac{1}{2})x \geq 2(-\frac{2}{1})$
$x \leq -4$

Always change the sign when multiplying or dividing by a negative number.

Solve each inequality and graph the answer on the number line.

1. $2.3 \geq s + 3$

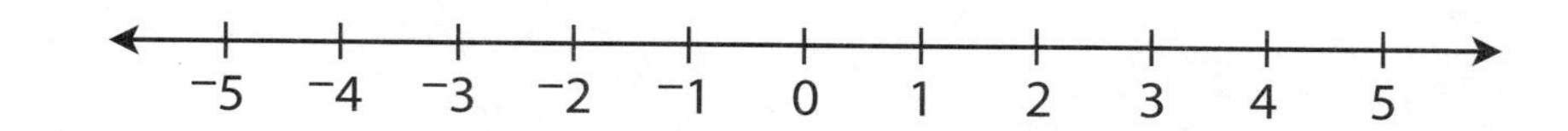

2. $6 > y + 2$

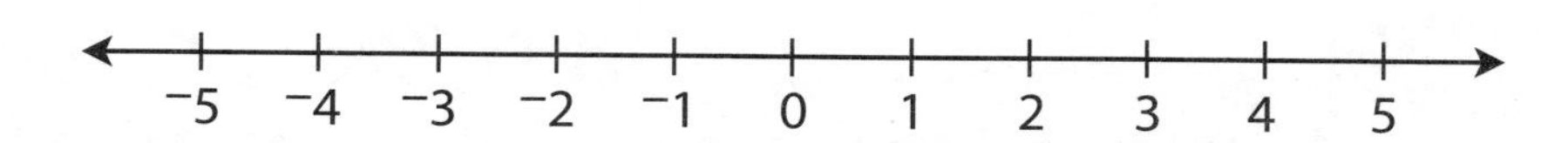

3. $7 + n \leq 9$

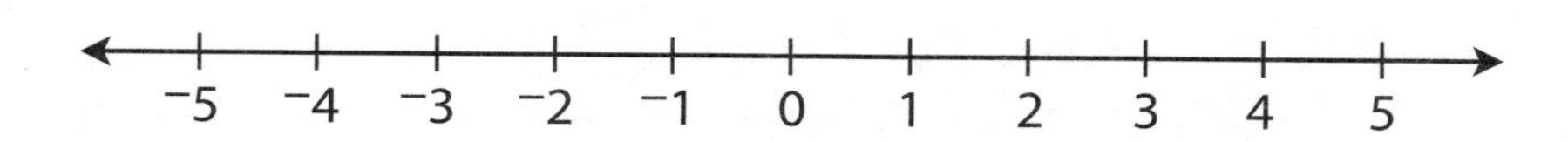

4. $d + \frac{2}{3} \geq \frac{1}{3}$

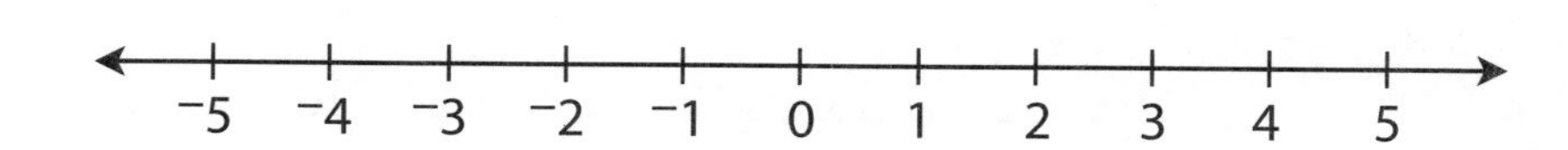

5. $-20n \leq -40$

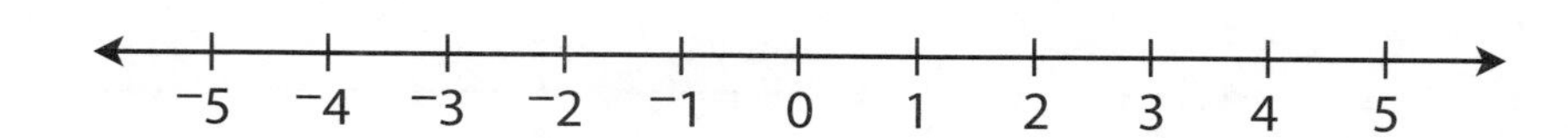

6. $12x > 24$

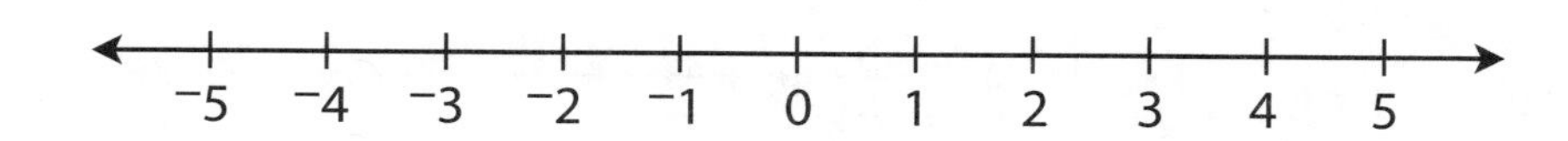

7. $-\frac{3}{4} \leq -3c$

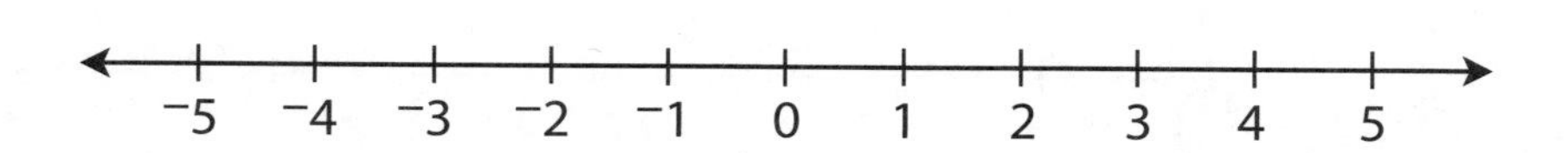

8. $4n \geq 2$

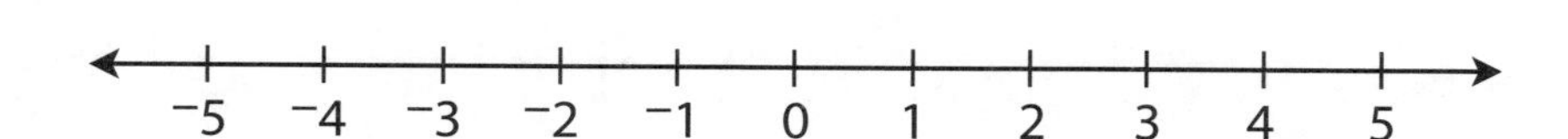

Name ______________________________ 7.EE.B.4b

Solving Inequalities

$$-20x + 8 > 4x - 40$$
$$-20x + 20x + 8 > 4x + 20x - 40$$
$$8 > 24x - 40$$
$$48 > 24x$$
$$2 > x$$

Solve each inequality and graph the answer on the number line.

1. $-6\frac{1}{3} \geq k + \frac{2}{3}$

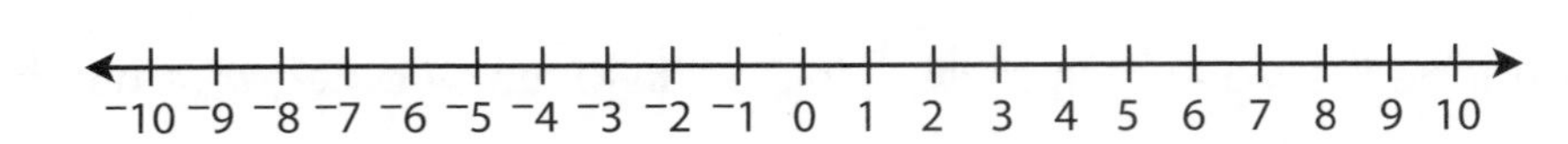

2. $h - 3 \leq 2$

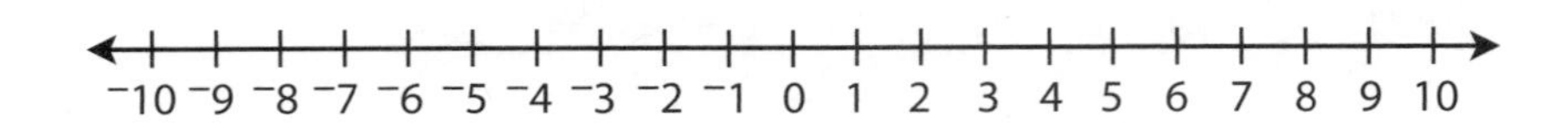

3. $a - 2.07 \geq 3.93$

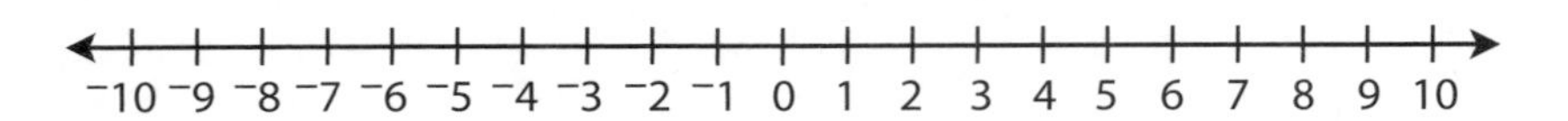

4. $1.3q \geq 8.7$

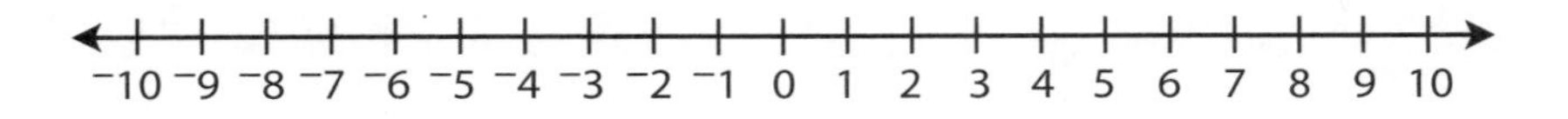

5. $-11 < k + -13$

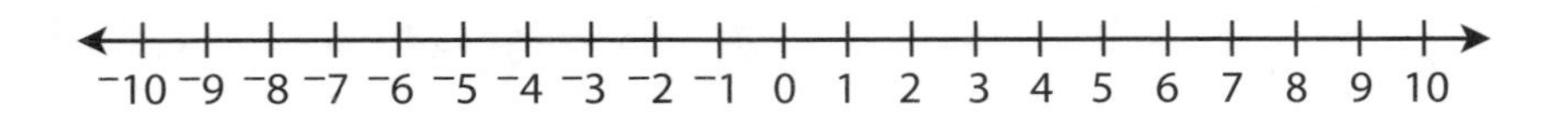

6. $-14 \geq h - 8$

−10 −9 −8 −7 −6 −5 −4 −3 −2 −1 0 1 2 3 4 5 6 7 8 9 10

7. $13 > 12m + 7$

−10 −9 −8 −7 −6 −5 −4 −3 −2 −1 0 1 2 3 4 5 6 7 8 9 10

8. $(\frac{2}{5})b \geq -2$

−10 −9 −8 −7 −6 −5 −4 −3 −2 −1 0 1 2 3 4 5 6 7 8 9 10

9. $-\frac{r}{2} \leq 5$

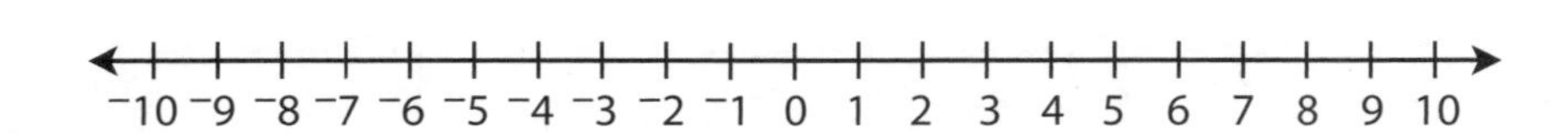

10. $-14d > 56$

−10 −9 −8 −7 −6 −5 −4 −3 −2 −1 0 1 2 3 4 5 6 7 8 9 10

Name ________________________________

7.EE.B.4b

Solving Inequalities

Solve each inequality and graph the answer on the number line.

1. $c + 1 > {}^{-}4$

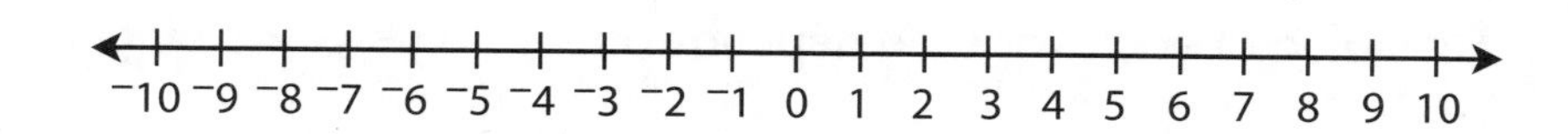

2. $6.5c < 6.5$

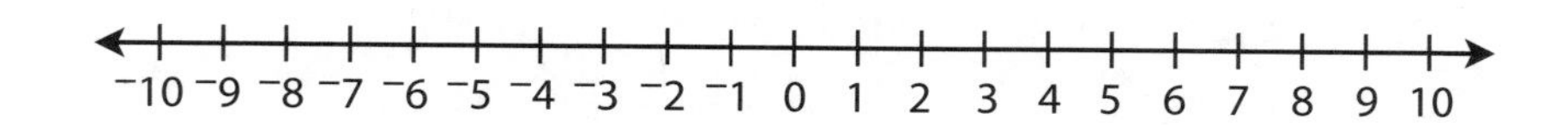

3. ${}^{-}4 \geq s - ({}^{-}2)$

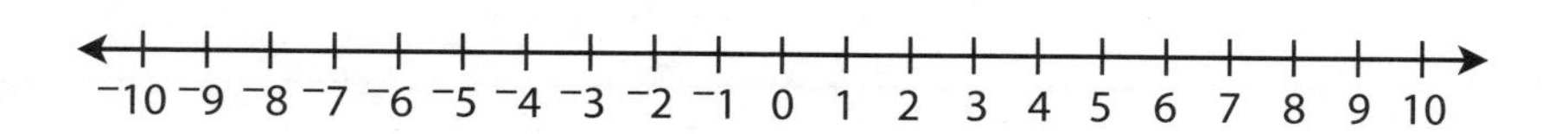

4. $(\frac{1}{2})y > {}^{-}5$

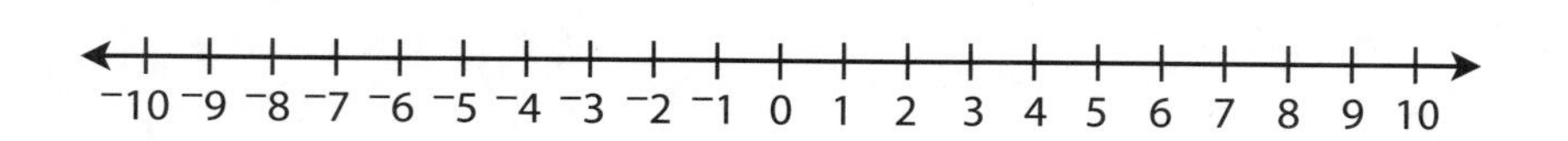

5. $d - 4.5 \geq {}^{-}1.5$

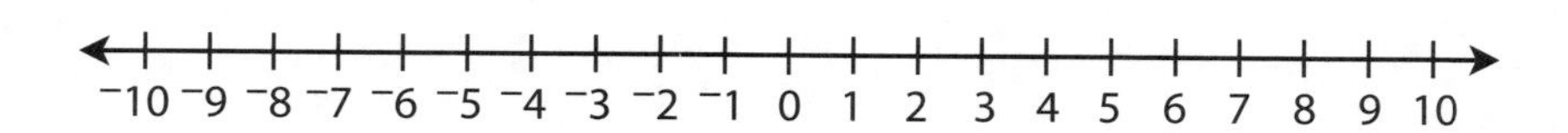

6. ${}^{-}14.5 \leq x + {}^{-}21.5$

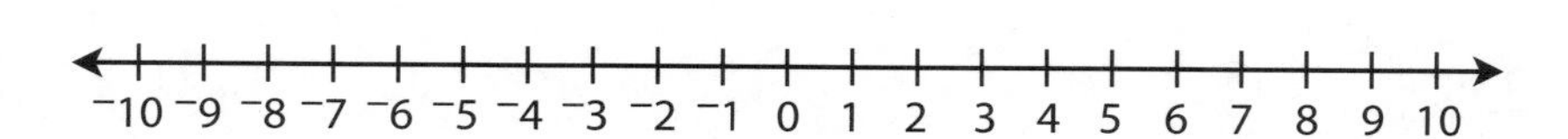

7. ${}^{-}13 < g - 12$

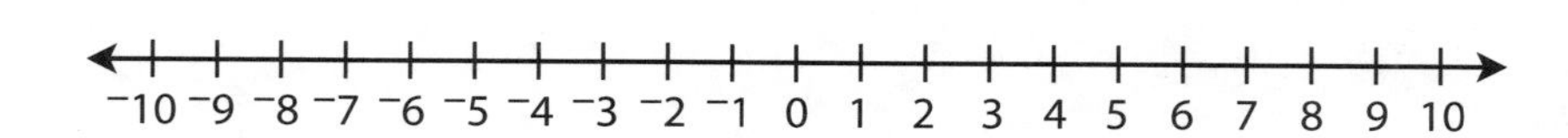

8. $h + 9 > 12$

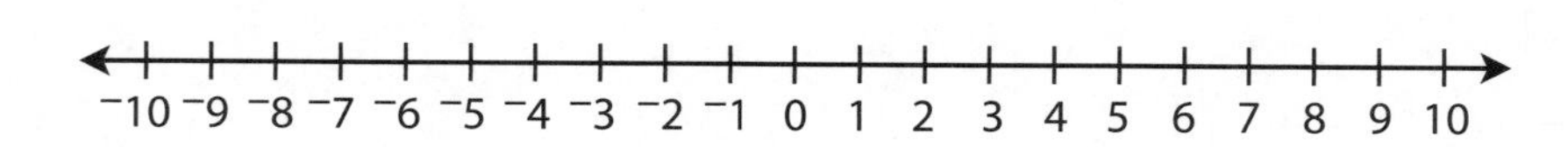

9. $-\frac{n}{3} \geq 2$

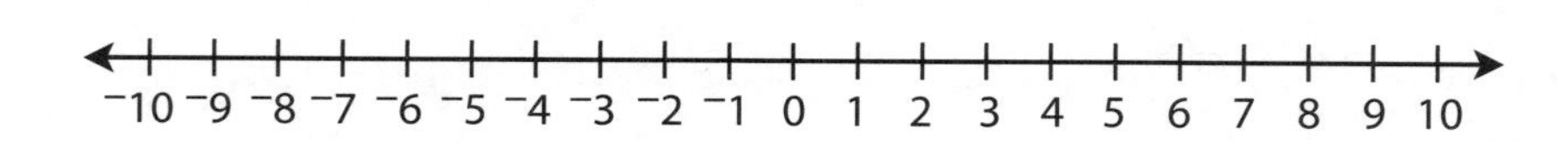

10. $10 > r + 14$

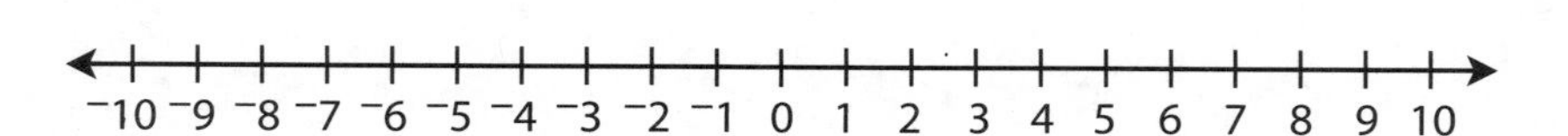

Name ______________________________ 6.EE.A.2c

Function Tables

When given the value of one variable in an equation, the value of the other variable can be found. The value of each variable depends on the value of the other.

$$y = 8x + 6$$

x	y
$^{-}1$	$^{-}2$
$^{-}2$	$^{-}10$
2	22

Create a function table for each equation using the given values of x.

1. $2 = y - 4x$
   let $x = {}^{-}1, 0, 2$

2. $^{-}x = y - 4$
   let $x = {}^{-}1, {}^{-}2, {}^{-}3$

3. $2x + y = {}^{-}5$
   let $x = {}^{-}3, 2, 4$

4. $2 = y - 3x$
   let $x = {}^{-}1, 0, 2$

5. $4 - y = 2x$
   let $x = {}^{-}3, 1, {}^{-}2$

6. $2x + y = 5$
   let $x = {}^{-}4, 0, 2$

7. $4x - y = {}^{-}10$
   let $x = 2, 3, {}^{-}2$

8. $4x - 2y = 6$
   let $x = {}^{-}1, \frac{1}{2}, 2$

Name ________________________________ 6.EE.A.2c

Function Tables

Remember, when given the value of one variable in an equation, the value of the other variable can be found. The value of each variable depends on the value of the other.

Create a function table for each equation using the given values of x.

1. $y + 3x = 7$
   let $x = {}^{-}9, {}^{-}5, 8$

2. $y = {}^{-}2x + 8$
   let $x = {}^{-}7, {}^{-}1, 2$

3. ${}^{-}7 - 6x = y$
   let $x = {}^{-}6, {}^{-}4, 9$

4. $y + 2 = {}^{-}5x,$
   let $x = {}^{-}4, {}^{-}1, 3$

5. $y = {}^{-}4x - 6$
   let $x = {}^{-}6, {}^{-}3, 0$

6. $y - 9x = {}^{-}5,$
   let $x = {}^{-}7, 3, 7$

7. $6 = y - 8x$
   let $x = {}^{-}5, 1, 8$

8. $y + 4 = {}^{-}9x$
   let $x = {}^{-}2, 0, 2$

9. $y = 5x + 3$
   let $x = {}^{-}2, 2, 4$

10. $2x = y - 2$
    let $x = {}^{-}3, 3, 6$

Name ______________________________ 6.EE.A.2c

Function Tables

Create a function table for each equation using the given values of x.

1. $y = {}^{-}2x - 3$
   let $x = {}^{-}8, 1, 6$

2. $y + \frac{1}{2}x = 3,$
   let $x = {}^{-}5, 1, 5$

3. $y + 1 = \frac{1}{2}x$
   let $x = {}^{-}7, 0, 7$

4. $y + \frac{1}{4}x = {}^{-}5$
   let $x = {}^{-}8, {}^{-}2, 4$

5. $y = {}^{-}3x - 4$
   let $x = {}^{-}4, 0, 3$

6. $\frac{1}{2}x - 2 = y$
   let $x = {}^{-}7, {}^{-}4, 4$

7. $5 + y = -\frac{1}{2}x$
   let $x = {}^{-}2, 6, 8$

8. $y = \frac{1}{4}x + 2$
   let $x = {}^{-}7, 0, 8$

9. $1 = -\frac{1}{2}x - y$
   let $x = {}^{-}6, 2, 3$

10. $y = -\frac{1}{4}x + 2$
    let $x = {}^{-}6, {}^{-}1, 3$

11. $y + \frac{1}{4}x = {}^{-}2$
    let $x = {}^{-}7, 0, 7$

12. $y + 2 = \frac{1}{4}x$
    let $x = {}^{-}8, {}^{-}5, {}^{-}4$

Name ______________________________ 5.G.A.1, 6.NS.C.6c

Plotting Ordered Pairs

(x,y) = (2, ⁻3) Right 2 and down 3

(x,y) = (⁻4, 1) Left 4 and up 1

(x,y) = (1, 3) Right 1 and up 3

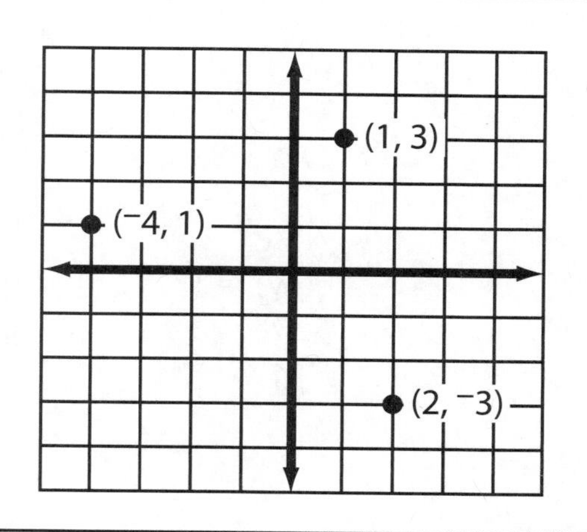

Plot and label the following points on the graph.

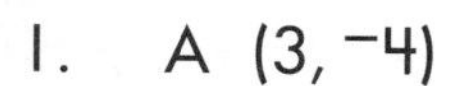

1. A (3, ⁻4)

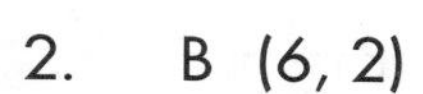

2. B (6, 2)
3. C (0, ⁻2)
4. D (1, 7)
5. E (3, ⁻3)
6. F (2, ⁻6)
7. G (⁻3, 4)
8. H (⁻1, ⁻4)
9. I (3, 0)
10. J (2, 5)

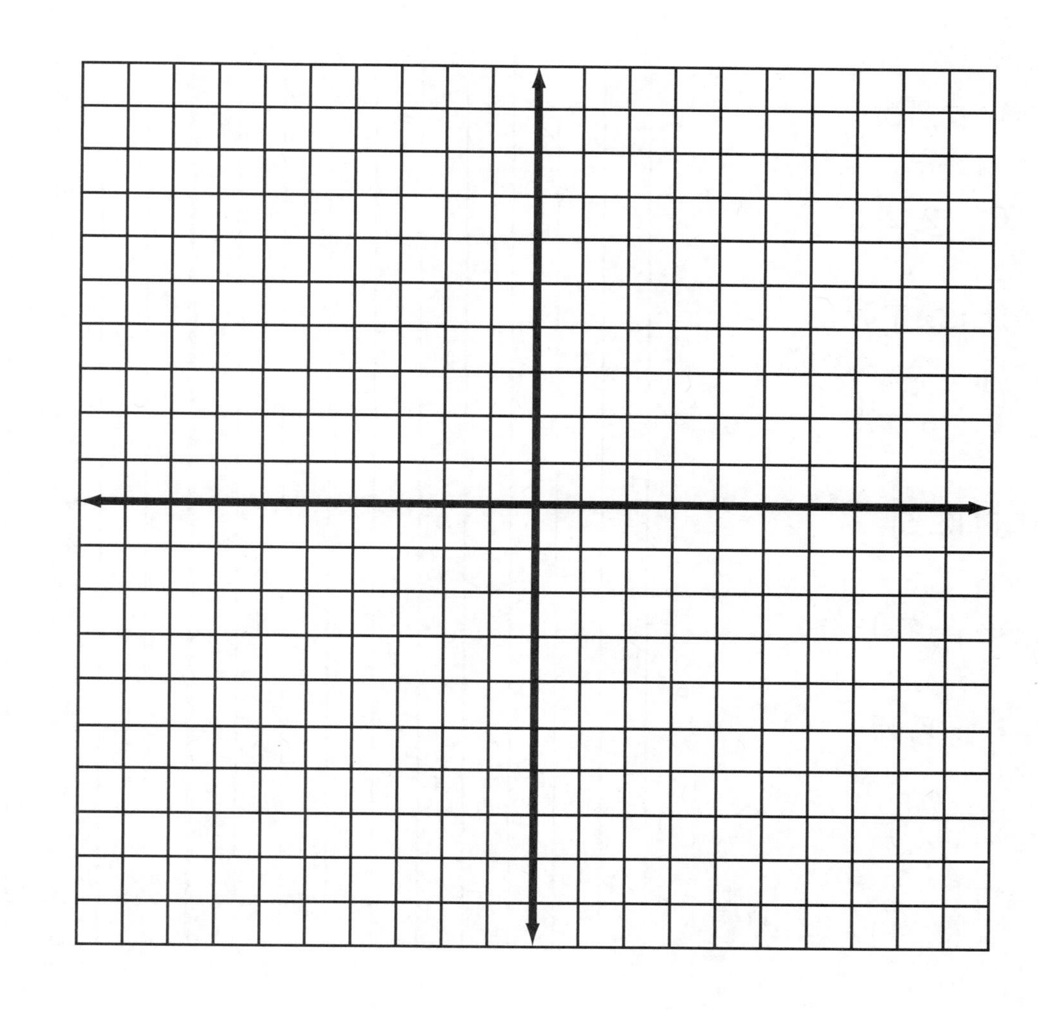

Name ____________________ 5.G.A.1, 6.NS.C.6c

Plotting Ordered Pairs

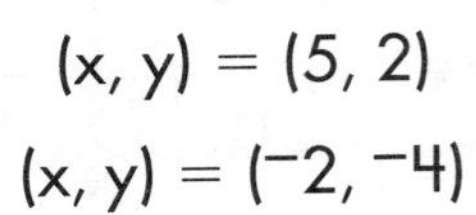

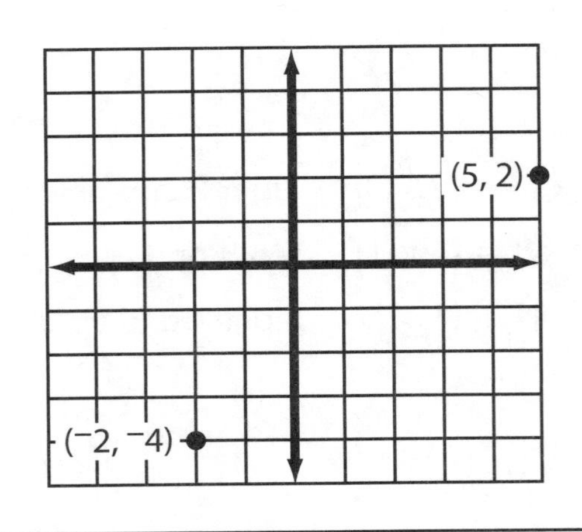

Plot and label the following points on the graph.

1. A (4, −2)
2. B (7, 10)
3. C (−2, 2)
4. D (3, 0)
5. E (−3, −8)
6. F (0, 5)
7. G (1, −2)
8. H (−7, 7)
9. I (5, 1)
10. J (6, −4)

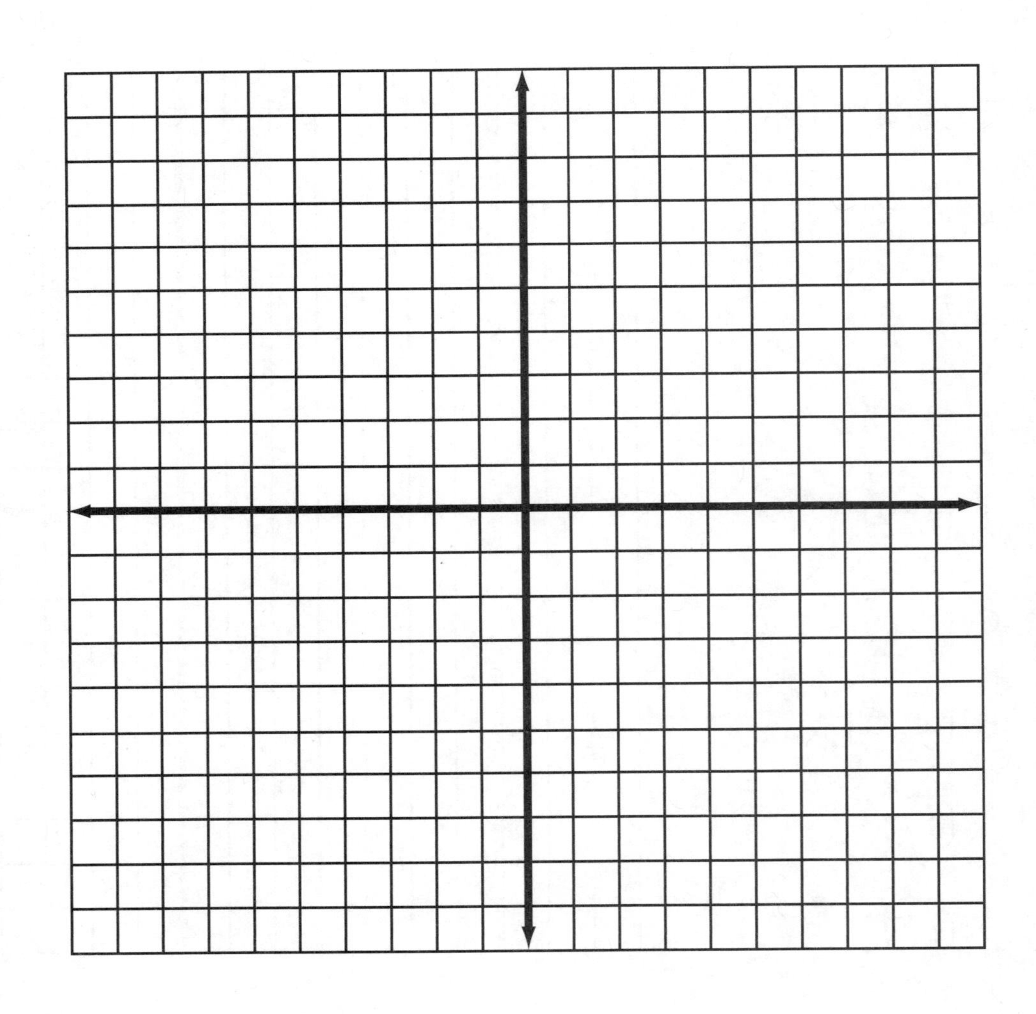

Name ______________________________

5.G.A.1, 6.NS.C.6c

Plotting Ordered Pairs

Plot and label the following points on the graph.

1. A (3, $^-$2)
2. B (7, 6)
3. C (0, $^-$3)
4. D ($^-$1, 4)
5. E ($^-$4, $^-$5)
6. F (7, 0)
7. G ($^-$1, 5)
8. H (3, $^-$4)
9. I (0, $^-$2)
10. J (6, 3)
11. K ($^-$4, $^-$1)
12. L ($^-$5, 0)

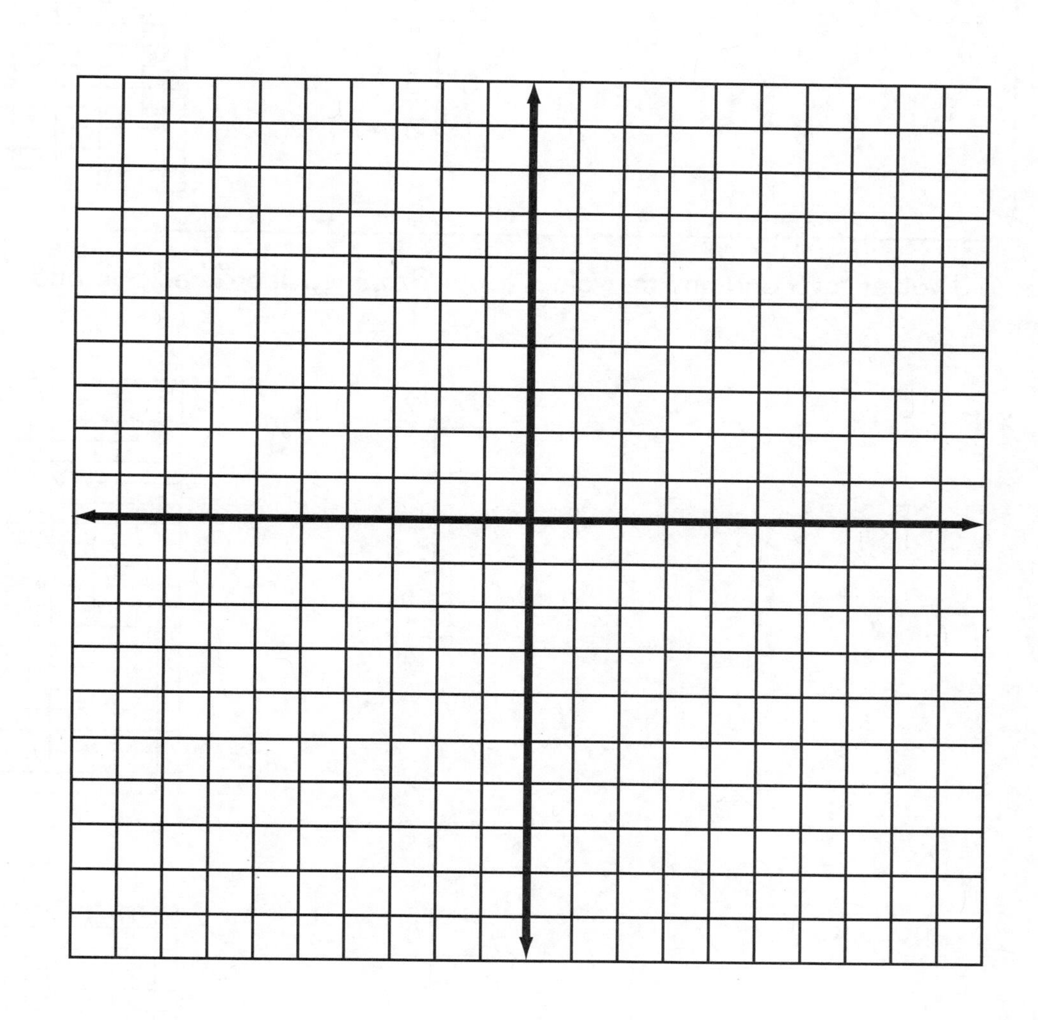

Name ______________________ 8.EE.C.7

Graphing Linear Equations

First, isolate the variable y. Then, choose several values for x and find the y value for each. Then, graph the coordinate pairs and draw a line connecting the points.

$x + 4 = y - 1$
$y = x + 5$

x	y
$^-1$	4
0	5
1	6

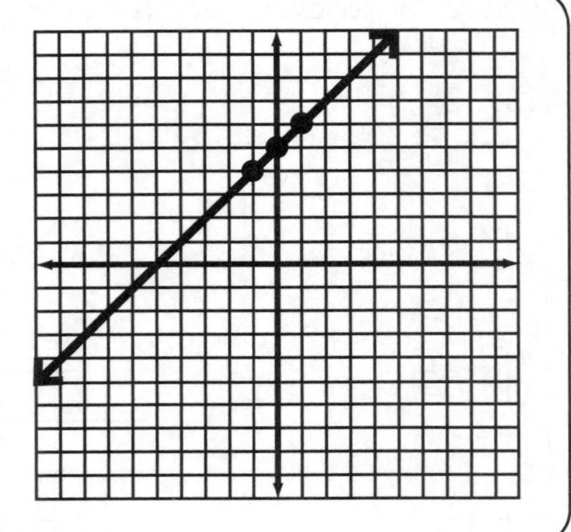

Choose 3 values for x and find the values for y. Graph each ordered pair and draw a line connecting them.

1. $y + 3 = 3x$

x	y

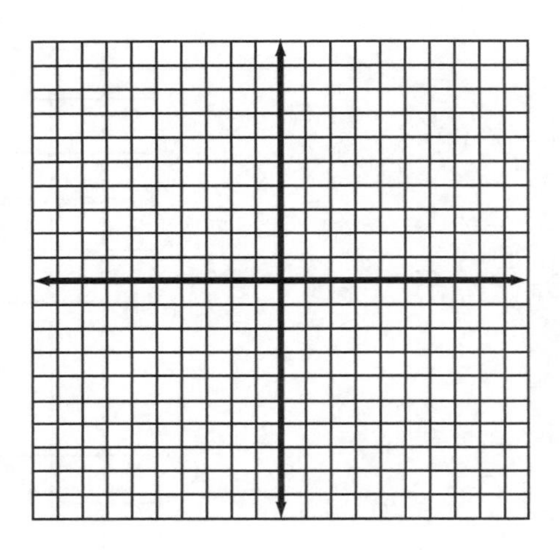

2. $2y = 4x + 4$

x	y

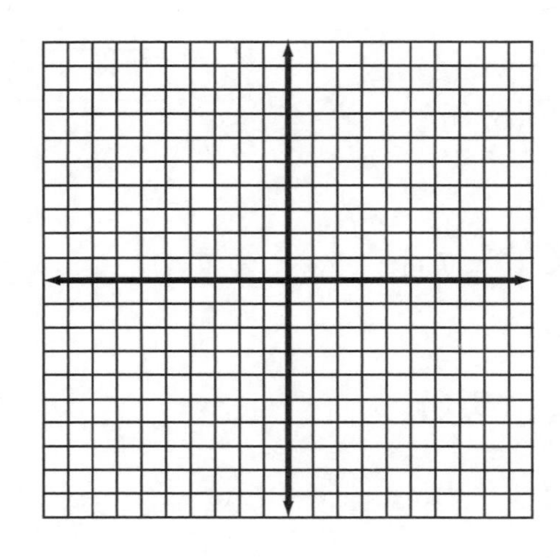

Name ____________________

8.EE.C.7

Graphing Linear Equations

Choose 3 values for x and find the values for y. Graph each ordered pair and draw a line connecting them.

1. $y - 4 = 2x$

x	y

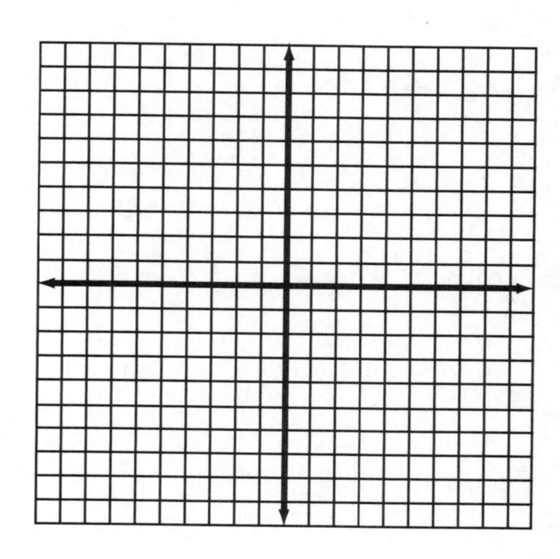

2. $y + 4 = 3x$

x	y

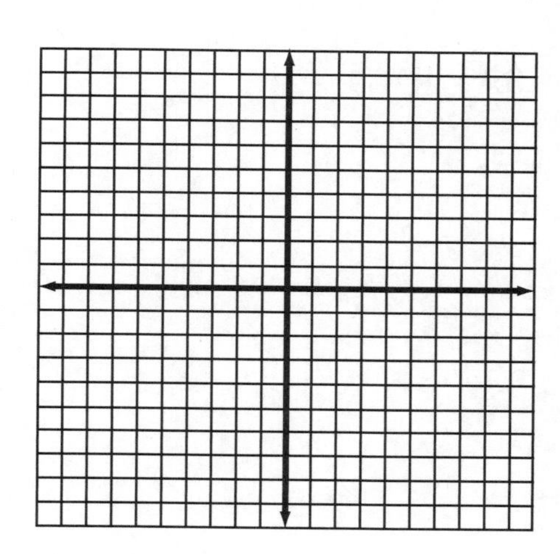

3. $4x - y = 8$

x	y

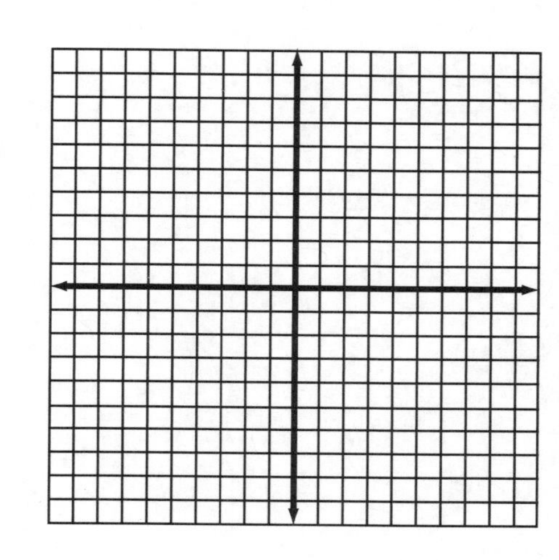

Name ______________________________ 8.EE.C.7

Graphing Linear Equations

Find 3 coordinate pairs for each equation. Graph each equation.

1. $4x - y = 5$

x	y

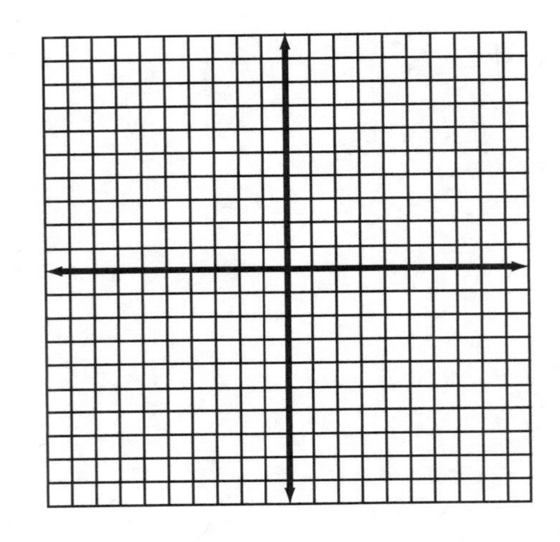

2. $y + 3x = 4$

x	y

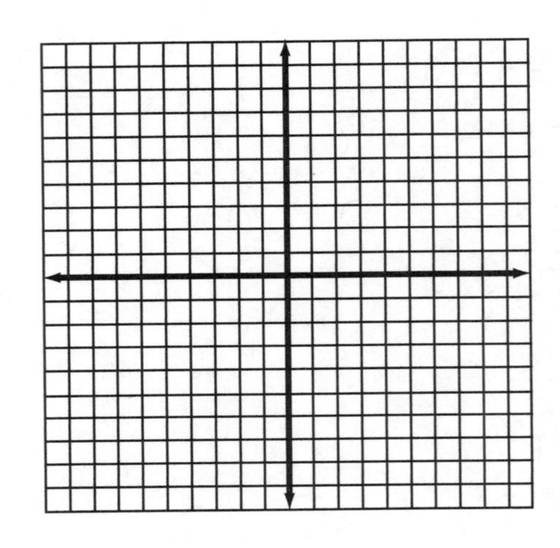

3. $3x - y = 5$

x	y

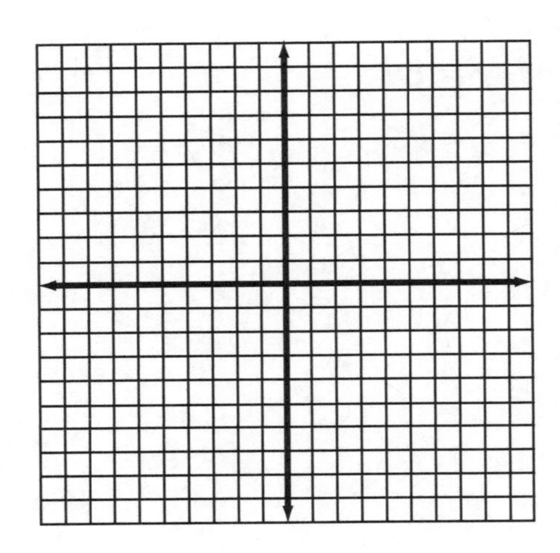

Name ______________________________ 8.EE.B.5, 8.F.B.4

Determining Slope

The **slope** of a line describes how steep it is. It is calculated by dividing the vertical rate of change (change in y) by the horizontal change (change in x). You can subtract the change in either order (A – B, or B – A), as long as both are subtracted in the same order.

A(⁻1, 4) B(3, 2)

slope (m) $= 2 - 4 \div 3 - (^{-}1) = \dfrac{2 - 4}{3 - (^{-}1)}$

$m = -\dfrac{1}{4}$

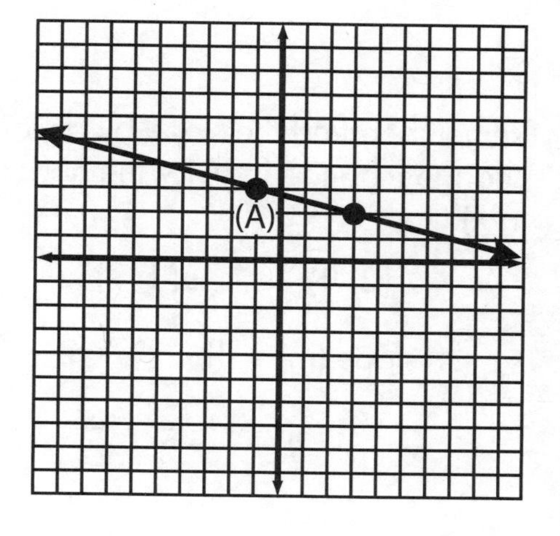

Find the slope of the line that contains each pair of points.

1. A(2, 4), B(4, 6)

2. A(3, 2), B(⁻2, ⁻8)

3. A(⁻1, 3), B(⁻2, 5)

4. A(8, ⁻3), B(10, 0)

5. A(0, 0), B(6, ⁻3)

6. A(3, 4), B(⁻1, 4)

7. A(⁻4, 7), B(⁻7, 15)

8. A(2, ⁻2), B(6, 5)

Name ______________________________ 8.EE.B.5, 8.F.B.4

Determining Slope

Remember, find the slope by dividing the vertical rate of change (change in y) by the horizontal change (change in x).

A(2, 4) B(4, 9)

slope (m) $= 4 - 9 \div 2 - 4 \quad \dfrac{4 - 9}{2 - 4}$

$m = -\dfrac{-5}{-2} = \dfrac{5}{2}$

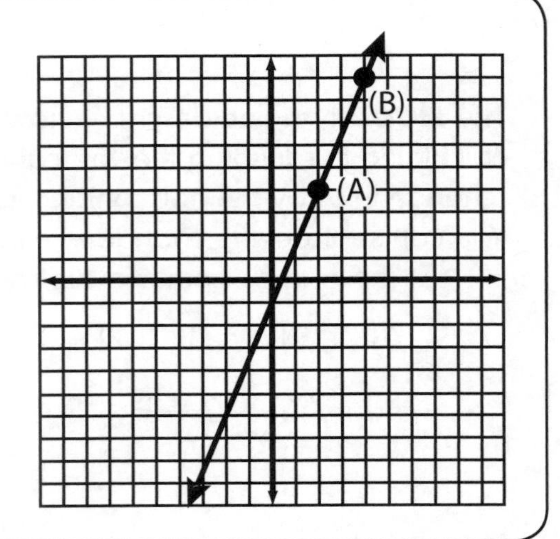

Find the slope of the line that contains each pair of points.

1. A(0, 1), B(2, 4)

2. A(2, 2), B(3, 3)

3. A(2, 0), B(3, 0)

4. A(3, 0), B(2, 3)

5. A($^-$1, 0), B($^-$3, 1)

6. A($^-$1, $^-$1), B($^-$3, 2)

7. A(1, 4), B($^-$1, 2)

8. A($^-$4, $^-$4), B(0, $^-$3)

9. A($^-$1, 2), B($^-$3, $^-$3)

10. A(0, 3), B(3, 4)

Name ______________________________

8.EE.B.5, 8.F.B.4

Determining Slope

Find the slope of the line that contains each pair of points.

1. A(0, ⁻2), B(1, 3)

2. A(2, ⁻1), B(5, ⁻2)

3. A(⁻3, 0), B(⁻2, 5)

4. A(⁻4, ⁻1), B(1, 3)

5. A(0, 1), B(⁻4, 4)

6. A(1, 0), B(4, 1)

7. A(⁻2, 0), B(0, ⁻2)

8. A(⁻1, ⁻2), B(1, 4)

9. A(⁻1, 2), B(⁻2, 5)

10. A(0, ⁻1), B(5, ⁻2)

11. A(⁻3, 1), B(2, ⁻2)

12. A(10, 3), B(7, 9)

Answer Key

Name ____________________ 6.NS.A.1

Dividing Fractions

rewrite — invert and multiply

$1\frac{2}{3} \div 2\frac{1}{5} = \frac{5}{3} \div \frac{11}{5} = \frac{5}{3} \times \frac{5}{11} = \frac{25}{33}$

rewrite

Solve each problem. Write the answer in simplest form.

1. $6\frac{2}{3} \div 4\frac{4}{9} =$ $\mathbf{1\frac{1}{2}}$
2. $3\frac{1}{3} \div 1\frac{5}{9} =$ $\mathbf{2\frac{1}{7}}$
3. $2\frac{7}{10} \div 3\frac{9}{15} =$ $\mathbf{\frac{3}{4}}$
4. $4\frac{1}{2} \div 5\frac{1}{4} =$ $\mathbf{\frac{6}{7}}$
5. $6\frac{3}{4} \div 2\frac{1}{2} =$ $\mathbf{2\frac{7}{10}}$
6. $2\frac{2}{6} \div 4\frac{2}{3} =$ $\mathbf{\frac{1}{2}}$
7. $5\frac{2}{5} \div 4\frac{1}{2} =$ $\mathbf{1\frac{1}{5}}$
8. $7\frac{2}{7} \div 2\frac{2}{14} =$ $\mathbf{3\frac{2}{5}}$
9. $3\frac{1}{2} \div 4\frac{1}{3} =$ $\mathbf{\frac{21}{26}}$
10. $2\frac{2}{3} \div 3\frac{4}{10} =$ $\mathbf{\frac{40}{51}}$
11. $4\frac{1}{5} \div 3\frac{3}{5} =$ $\mathbf{1\frac{1}{6}}$
12. $5\frac{3}{5} \div 1\frac{5}{9} =$ $\mathbf{3\frac{3}{5}}$

Name ____________________ 6.NS.A.1

Dividing Fractions

rewrite — invert and multiply

$1\frac{1}{8} \div 2\frac{1}{6} = \frac{9}{8} \div \frac{13}{6} = \frac{9}{8} \times \frac{6}{13} = \frac{27}{52}$

rewrite

Solve each problem. Write the answer in simplest form.

1. $9\frac{1}{6} \div 3\frac{5}{12} =$ $\mathbf{2\frac{28}{41}}$
2. $9\frac{1}{6} \div 3\frac{8}{12} =$ $\mathbf{2\frac{1}{2}}$
3. $7\frac{1}{2} \div 8\frac{3}{4} =$ $\mathbf{\frac{6}{7}}$
4. $5\frac{1}{2} \div 8\frac{4}{5} =$ $\mathbf{\frac{5}{8}}$
5. $5\frac{4}{5} \div 1\frac{8}{15} =$ $\mathbf{3\frac{18}{23}}$
6. $9\frac{1}{5} \div 2\frac{3}{10} =$ $\mathbf{4}$
7. $7\frac{4}{5} \div 1\frac{3}{10} =$ $\mathbf{6}$
8. $7\frac{1}{9} \div 2\frac{2}{3} =$ $\mathbf{2\frac{2}{3}}$
9. $8\frac{4}{5} \div 1\frac{1}{15} =$ $\mathbf{8\frac{1}{4}}$
10. $8\frac{2}{5} \div 2\frac{1}{10} =$ $\mathbf{4}$
11. $5\frac{3}{5} \div 1\frac{6}{10} =$ $\mathbf{3\frac{1}{2}}$
12. $6\frac{1}{3} \div 2\frac{1}{6} =$ $\mathbf{2\frac{12}{13}}$
13. $11\frac{3}{4} \div 5\frac{1}{2} =$ $\mathbf{2\frac{3}{22}}$
14. $8\frac{3}{5} \div 2\frac{7}{10} =$ $\mathbf{3\frac{5}{27}}$
15. $3\frac{5}{7} \div 3\frac{13}{14} =$ $\mathbf{\frac{52}{55}}$

Name ____________________ 6.NS.A.1

Dividing Fractions

Solve each problem. Write the answer in the simplest form.

1. $\frac{1}{4} \div 4 =$ $\mathbf{\frac{1}{16}}$
2. $1\frac{1}{6} \div 2\frac{1}{2} =$ $\mathbf{\frac{7}{15}}$
3. $\frac{5}{6} \div \frac{2}{3} =$ $\mathbf{1\frac{1}{4}}$
4. $\frac{1}{5} \div \frac{1}{6} =$ $\mathbf{1\frac{1}{5}}$
5. $\frac{3}{5} \div \frac{7}{10} =$ $\mathbf{\frac{6}{7}}$
6. $2\frac{4}{7} \div \frac{1}{8} =$ $\mathbf{20\frac{4}{7}}$
7. $3\frac{1}{3} \div 5\frac{1}{2} =$ $\mathbf{\frac{20}{33}}$
8. $3\frac{1}{5} \div 1\frac{6}{10} =$ $\mathbf{2}$
9. $2\frac{2}{9} \div 4\frac{1}{6} =$ $\mathbf{\frac{8}{15}}$
10. $4\frac{3}{5} \div 1\frac{3}{8} =$ $\mathbf{3\frac{19}{55}}$
11. $3\frac{3}{4} \div 3\frac{1}{8} =$ $\mathbf{1\frac{1}{5}}$
12. $9\frac{3}{7} \div 5\frac{10}{14} =$ $\mathbf{1\frac{13}{20}}$
13. $5\frac{1}{6} \div 2\frac{7}{12} =$ $\mathbf{2}$
14. $\frac{8}{7} \div 4 =$ $\mathbf{\frac{2}{7}}$
15. $1\frac{1}{2} \div 2\frac{1}{4} =$ $\mathbf{\frac{2}{3}}$
16. $6\frac{1}{5} \div 2\frac{2}{3} =$ $\mathbf{2\frac{13}{40}}$
17. $1\frac{2}{3} \div 1\frac{1}{3} =$ $\mathbf{1\frac{1}{4}}$
18. $1\frac{2}{5} \div 2\frac{3}{4} =$ $\mathbf{\frac{28}{55}}$
19. $3\frac{1}{9} \div 12\frac{2}{8} =$ $\mathbf{\frac{16}{63}}$
20. $3\frac{4}{10} \div 3\frac{6}{7} =$ $\mathbf{\frac{119}{135}}$
21. $4\frac{5}{9} \div 3\frac{5}{12} =$ $\mathbf{1\frac{1}{3}}$

Name ____________________ 5.NBT.B.7, 6.NS.B.3

Adding Decimals

$$\begin{array}{r} 12.2 + 5.25 = \quad 12.20 \\ +\ 5.25 \\ \hline 17.45 \end{array}$$

Solve each problem.

1. 9.87 + 2.87 = **12.74**
2. 5.02 + 8.2 = **13.22**
3. 2.49 + 4.73 = **7.22**
4. 6.41 + 2.734 + 8.41 = **17.554**
5. 2.934 + 231.6 = **234.534**
6. 121.9 + 0.736 = **122.636**
7. 43.56 + 85.7 = **129.26**
8. 13.238 + 4.82 = **18.058**
9. 15.76 + 25.23 + 3.9 = **44.89**
10. 6.41 + 3.99 = **10.4**
11. 6.3 + 9.124 + 2.34 = **17.764**
12. 5.97 + 4.87 + 3.908 = **14.748**
13. 13.39 + 7.4 = **20.79**
14. 4.63 + 23.5 + 5.0 = **33.13**
15. 3.456 + 2.894 = **6.35**
16. 3.64 + 5.32 = **8.96**
17. 5.7 + 5.34 + 4.78 = **15.82**
18. 3.5 + 8.4 = **11.9**
19. 0.034 + 10.51 = **10.544**
20. 123.415 + 6.876 = **130.291**

Answer Key

Name

(5.NBT.B.7, 6.NS.B.3)

Adding Decimals

Remember:
To add decimals, align the numbers on the decimal. Add zeros as place holders. Then, add.

Solve each problem.

1. 40.14 + 12.53 + 5.6 = **58.27**
2. 3.43 + 5.45 = **8.88**
3. 3.59 + 2.08 = **5.67**
4. 17.34 + 6.45 = **23.79**
5. 2.15 + 4.25 = **6.4**
6. 108.7 + 0.489 = **109.189**
7. 31.71 + 324.95 = **356.66**
8. 121.356 + 80.52 = **201.876**
9. 5.37 + 7.37 = **12.74**
10. 7.22 + 3.41 = **10.63**
11. 2.6 + 45.54 + 3.65 = **51.79**
12. 6.29 + 8.83 + 6.332 = **21.452**
13. 5.4 + 7.38 + 6.21 = **18.99**
14. 8.45 + 23.20 + 5.34 = **36.99**
15. 5.44 + 3.34 + 6.30 = **15.08**
16. 2.312 + 5.371 = **7.683**
17. 12.52 + 8.32 = **20.84**
18. 321.595 + 3.45 = **325.045**
19. 0.012 + 25.08 = **25.092**
20. 17.121 + 5.34 = **22.461**

Name

(5.NBT.B.7, 6.NS.B.3)

Adding Decimals

Solve each problem.

1. 42.25 + 53.5 = **95.75**
2. 12.828 + 10.548 = **23.376**
3. 0.6 + 0.09 + 1.75 = **2.44**
4. 65.78 + 54.90 = **120.68**
5. 6.77 + 0.05 = **6.82**
6. 20.59 + 44.5 = **65.09**
7. 154.34 + 42.98 = **197.32**
8. 17.546 + 5.0958 = **22.6418**
9. 23.565 + 28.403 = **51.968**
10. 0.8 + 0.07 + 3.73 = **4.6**
11. 4.35 + 5.4 = **9.75**
12. 52.4 + 0.62 + 2.096 = **55.116**
13. 16.08 + 3.2 = **19.28**
14. 22.3 + 0.89 + 27 = **50.19**
15. 7.150 + 0.263 = **7.413**
16. 0.258 + 32.900 = **33.158**
17. 75.10 + 3.62 = **78.72**
18. 13.80 + 6.32 = **20.12**
19. 0.894 + 1.773 = **2.667**
20. 12.354 + 15.438 = **27.792**

Name

(5.NBT.B.7, 6.NS.B.3)

Subtracting Decimals

13.2 − 4.10 = 13.20 − 4.10 = 9.10

Solve each problem.

1. 4.239 − 0.06 = **4.179**
2. 51.23 − 14.45 = **36.78**
3. 16.3 − 12.4 = **3.9**
4. 452.82 − 127.36 = **325.46**
5. 62.1 − 33.29 = **28.81**
6. 75.034 − 22.439 = **52.595**
7. 76.34 − 47.30 = **29.04**
8. 34.32 − 12.43 = **21.89**
9. 435.34 − 345.34 = **90**
10. 756.98 − 32.43 = **724.55**
11. 513.43 − 305.342 = **208.088**
12. 65.9 − 33.2 = **32.7**
13. 21.73 − 16.43 = **5.3**
14. 121.32 − 19.34 = **101.98**
15. 23.28 − 0.552 − 1.2 = **21.528**
16. 8.64 − 0.476 = **8.164**
17. 13.2 − 6.7 = **6.5**
18. 21.32 − 4.28 = **17.04**
19. 35.63 − 0.021 = **35.609**
20. 485.02 − 332.86 = **152.16**

Name

(5.NBT.B.7, 6.NS.B.3)

Subtracting Decimals

Remember:
To subtract decimals, align the numbers on the decimal. Add zeros as place holders. Then, subtract.

Solve each problem.

1. 43,289.56 − 28,125.87 = **15,163.69**
2. 756.84 − 31.343 = **725.497**
3. 34.34 − 23.19 = **11.15**
4. 4.7 − 2.3 = **2.4**
5. 95.87 − 52.45 = **43.42**
6. 72.72 − 43.562 = **29.158**
7. 85.76 − 34.65 = **51.11**
8. 7.435 − 0.0345 = **7.4005**
9. 345.24 − 159.24 = **186**
10. 54.68 − 23.76 = **30.92**
11. 74.71 − 61.92 = **12.79**
12. 84.8 − 44.87 = **39.93**
13. 857.44 − 22.39 = **835.05**
14. 93.76 − 8.67 = **85.09**
15. 233.23 − 6.45 = **226.78**
16. 6.56 − 0.654 = **5.906**
17. 43.5 − 0.015 − 3.2 = **40.285**
18. 39.43 − 15.34 = **24.09**
19. 56.4 − 0.043 = **56.357**
20. 954.34 − 657.56 = **296.78**

Answer Key

Name ______________________ (5.NBT.B.7, 6.NS.B.3)

Subtracting Decimals

Solve each problem.

1. 725.987 − 231.155 = **494.832**
2. 13.58 − 7.2 = **6.38**
3. 432.42 − 327.89 = **104.53**
4. 1,343.32 − 1,032.90 = **310.42**
5. 21.7 − 15.9 = **5.8**
6. 843.21 − 452.03 = **391.18**
7. 15.51 − 8.34 = **7.17**
8. 545.825 − 137.405 = **408.42**
9. 43.5 − 14.2 = **29.3**
10. 1,350.65 − 253.42 = **1,097.23**
11. 2.55 − 2.05 = **0.5**
12. 5.085 − 2.5 = **2.585**
13. 12 − 2.73 = **9.27**
14. 26.8 − 6.48 = **20.32**
15. 2.04 − 0.998 = **1.042**
16. 66.5 − 0.675 = **65.825**
17. 35.91 − 17.16 = **18.75**
18. 0.9505 − 0.02709 = **0.92341**
19. 1 − 0.98765 = **0.01235**
20. 2.35 − 0.9 = **1.45**

Name ______________________ (5.NBT.B.7, 6.NS.B.3)

Multiplying Decimals

(0.3)(0.12) → 3 decimal places → 0.3 × 0.12 = 0.036

Solve each problem. Show your work.

1. (0.6)(0.022) = **0.0132**
2. (0.012)(0.7) = **0.0084**
3. (3.2)(0.65) = **2.08**
4. 0.07 × 0.4 = **0.028**
5. (0.02)(1.2) = **0.024**
6. 0.03 × 0.7 = **0.021**
7. (0.5)(0.2) = **0.1**
8. (2.2)(0.22) = **0.484**
9. (0.12)(0.04) = **0.0048**
10. 0.06 × 0.07 = **0.0042**
11. (0.11)(0.07) = **0.0077**
12. (0.13)(0.02) = **0.0026**
13. (0.7)(0.07) = **0.049**
14. (0.5)(0.05) = **0.025**
15. 0.5 × 0.06 = **0.03**
16. (0.012)(1.2) = **0.0144**
17. (0.8)(0.005) = **0.004**
18. 0.25 × 0.07 = **0.0175**
19. (0.9)(0.002) = **0.0018**
20. (0.9)(0.9) = **0.81**

Name ______________________ (5.NBT.B.7, 6.NS.B.3)

Multiplying Decimals

Remember:
To multiply decimals, align the numbers on the right side. Then, multiply. Place the decimal point in the product so it has the same number of decimal places as the factors.

Solve each problem. Show your work.

1. (7.8)(1.03) = **8.034**
2. (3.2)(3.065) = **9.808**
3. (12.2)(34.9) = **425.78**
4. (0.04)(0.24)(1.4) = **0.01344**
5. (2.5)(3.3)(0.33) = **2.7225**
6. (1.3)(3.04)(5.46) = **21.57792**
7. (4.3)(3.59) = **15.437**
8. (5.5)(4.304) = **23.672**
9. (23.4)(3.9) = **91.26**
10. (5.5)(2.6)(4.0) = **57.2**
11. (7)(20.2) = **141.4**
12. (6.2)(0.35) = **2.17**
13. (0.2)(0.18) = **0.036**
14. (8.5)(9.1) = **77.35**
15. (4.1)(4.1) = **16.81**
16. (5.014)(5.4) = **27.0756**
17. (7.5)(0.75) = **5.625**
18. (5.75)(1.2) = **6.9**
19. (65.7)(2.5) = **164.25**
20. (8.2)(0.2) = **1.64**

Name ______________________ (5.NBT.B.7, 6.NS.B.3)

Multiplying Decimals

Solve each problem. Show your work.

1. (12.3)(5.81)(0.06) = **4.28778**
2. (0.042)(0.006) = **0.000252**
3. (0.34)(0.12)(0.104) = **0.0042432**
4. (8.9)(0.11)(3.09) = **3.02511**
5. (15.92)(0.4)(0.32) = **2.03776**
6. (0.004)(6) = **0.024**
7. (6.4)(0.3) = **1.92**
8. (0.4)(0.232) = **0.0928**
9. (5.12)(6) = **30.72**
10. (10.89)(0.221) = **2.40669**
11. (3.28)(12.8) = **41.984**
12. (0.004)(0.0004)(0.04) = **0.000000064**
13. (0.016)(3.8) = **0.0608**
14. (0.007)(0.6)(0.05) = **0.00021**
15. (3.806)(10.01) = **38.09806**
16. (340)(0.02) = **6.8**
17. (0.8)(0.342)(0.02) = **0.005472**
18. (2.09)(0.005) = **0.01045**
19. (0.05)(0.15)(0.002) = **0.000015**
20. (5.4)(0.645)(0.07) = **0.24381**

Answer Key

Name ________________ (5.NBT.B.7, 6.NS.B.3)

Dividing Decimals

$1.2\overline{)0.84}$ Move the decimal points to make the divisor a whole number.

$12.\overline{)08.4}$ Divide. Add a decimal point to the quotient.

Solve each problem. Use mental math.

1. 0.036 ÷ 0.6 = **0.06**
2. 0.55 ÷ 0.005 = **110**
3. 7.2 ÷ 1.2 = **6**
4. 100 ÷ 0.01 = **10000**
5. 4.8 ÷ 0.06 = **80**
6. 0.0027 ÷ 0.9 = **0.003**
7. 1.69 ÷ 0.13 = **13**
8. 0.108 ÷ 0.09 = **1.2**
9. 0.44 ÷ 0.4 = **1.1**
10. 1.21 ÷ .11 = **11**
11. 8.4 ÷ 0.12 = **70**
12. 0.064 ÷ 0.8 = **0.08**
13. 3.6 ÷ 0.009 = **400**
14. 0.0054 ÷ 0.006 = **0.9**
15. 0.012 ÷ 0.3 = **0.04**
16. 14.4 ÷ 1.2 = **12**
17. 0.56 ÷ 0.008 = **70**
18. 2.6 ÷ 0.02 = **130**

Name ________________ (5.NBT.B.7, 6.NS.B.3)

Dividing Decimals

Remember:
To divide decimals, move the decimal points to make the divisor a whole number.
Then, divide. Add a decimal to the quotient.

Solve each problem. Be sure to mark repeating digits by placing a line above the digits that repeat.

1. 2.34 ÷ 1.6 = **1.4625**
2. 2.62 ÷ 0.54 = $\mathbf{4.\overline{851}}$
3. 23.65 ÷ 22.5 = $\mathbf{1.05\overline{1}}$
4. 872.6 ÷ 2.4 = $\mathbf{363.58\overline{3}}$
5. 1.32 ÷ 1.8 = $\mathbf{0.7\overline{3}}$
6. 3.6 ÷ 0.3 = **12**
7. 44.34 ÷ 32.76 = $\mathbf{1.3\overline{534798}}$
8. 34.96 ÷ 3.549 = **9.850662158**
9. 7.569 ÷ 3.459 = **2.188204683**
10. 8.37 ÷ 4.50 = **1.86**
11. 9.2 ÷ 1.2 = $\mathbf{7.\overline{6}}$
12. 16 ÷ 0.48 = $\mathbf{33.\overline{3}}$
13. 0.014 ÷ 0.005 = **2.8**
14. 10.81 ÷ 9.1 = **1.18791209**
15. 16.25 ÷ 0.5 = **32.5**
16. 0.564 ÷ 1.2 = **0.47**
17. 8.704 ÷ 3.4 = **2.56**
18. 4.848 ÷ 0.08 = **60.6**
19. 0.0448 ÷ 0.014 = **3.2**
20. 10.98 ÷ 0.02 = **549**

Name ________________ (5.NBT.B.7, 6.NS.B.3)

Dividing Decimals

Solve each problem.

1. 6.3056 ÷ 4.2 = $\mathbf{1.501\overline{3}}$
2. 7.57 ÷ 0.1 = **75.7**
3. 3.56 ÷ 2.5 = **1.424**
4. 0.493 ÷ 0.33 = $\mathbf{1.\overline{493}}$
5. 8.565 ÷ 2 = **4.2825**
6. 0.0135 ÷ 4.5 = **0.003**
7. 40.78 ÷ 0.2 = **203.9**
8. 9.51 ÷ 3.03 = $\mathbf{3.\overline{1386}}$
9. 12.63 ÷ 0.9 = $\mathbf{14.0\overline{3}}$
10. 9.414 ÷ 3.3 = $\mathbf{2.85\overline{27}}$
11. 1.35 ÷ 0.07 = $\mathbf{19.\overline{285714}}$
12. 16.73 ÷ 0.12 = $\mathbf{139.41\overline{6}}$
13. 3.605 ÷ 3.2 = **1.1265625**
14. 0.1827 ÷ 0.09 = **2.03**
15. 12.264 ÷ 5.6 = **2.19**
16. 6.65 ÷ 2.4 = $\mathbf{2.7708\overline{3}}$
17. 2.34 ÷ 0.012 = **195**
18. 0.576 ÷ 4.1 = $\mathbf{0.14\overline{04878}}$
19. 15.8 ÷ 0.09 = $\mathbf{175.\overline{5}}$
20. 8.176 ÷ 3.2 = **2.555**
21. 0.0224 ÷ 3.6 = $\mathbf{0.006\overline{2}}$
22. 21.5 ÷ 0.05 = **430**

Name ________________ (5.NBT.B.7 6.NS.B.3)

Problem Solving with Decimals

John and Jason went to a grocery store and bought some sandwiches for $4.95, a gallon of fruit juice for $3.31, and a bag of carrots for $3.15. How much did they spend altogether?

$4.95 + $3.31 + $3.15 =

```
   $4.95
   $3.31
 + $3.15
  $11.41  total
```

Solve each problem. Show your work.

1. George went to the store to buy a pair of pants. The pair of pants that George picked out cost $45.00. If the price of the pants was reduced by $10.85, how much will George pay?

   $34.15

2. Gary spent $13.41 on clothes in January, $25.95 in February, and $31.50 in March. Altogether, how much money did he spend on clothes in these months?

   $70.86

3. Jesse buys a shirt for $32.95 and a pair of shoes for $46.25. How much money does Jesse spend in all?

   $79.20

4. Heather bought some new fishing equipment. She bought a tackle box for $24.95, a fishing pole for $58.49, a life jacket for $37.75, and an ice chest for $57.41. How much money did Heather spend on her equipment?

   $178.60

5. Marisol and Kurt are making sandwiches. The ingredients for the sandwiches cost $6.07. The sandwiches from the store cost $8.55. How much money are they saving?

   $2.48

6. Cindy buys a shirt for $26.70, a pair of jeans for $47.55, a jacket for $25.54, and a hat for $20.11. How much does Cindy spend in all?

   $119.90

Answer Key

Name____________________ (5.NBT.B.7, 6.NS.B.3)

Problem Solving with Decimals

Solve each problem. Show your work.

1. Ellen wanted to buy the following items: a DVD player for $49.95, a DVD holder for $19.95, and a personal stereo for $21.95. Does Ellen have enough money to buy all three items if she has $90 with her?

   No, she needs $91.85.

2. John bought a case of soda (24 cans) at the store for $0.31 a can. How much money did John spend at the store?

   $7.44

3. On Juan's road trip, he spent $15.50 for 12.5 gallons of gas. How much money did he pay per gallon?

   $1.24 per gallon

4. If a car uses 16.4 gallons of gas in 4 hours, how many gallons are used per hour?

   4.1 gallons per hour

5. Rob rode his bike 48.3 miles in 3 hours. How many miles did he bike in 1 hour?

   16.1 miles in 1 hour

6. Melissa purchased $39.46 in groceries at a store. Melissa paid with a $50 bill. How much change should the cashier give Melissa?

   $10.54

Name____________________ (5.NBT.B.7, 6.NS.B.3)

Problem Solving with Decimals

Solve each problem. Show your work.

1. A cylinder is normally 3.45 cm in diameter. If it is remade to be 0.056 cm larger, what is the new size of the cylinder?

   3.506 cm in diameter

2. Norma has $435.82 in her checking account. How much does she have in her account after she makes a deposit of $115.75 and a withdrawal of $175.90?

   $375.67

3. Alan is connecting three garden hoses to make one longer hose. One hose is 6.25 feet long, another hose is 5.755 feet long, and the last hose is 6.5 feet long. How long is the hose after they are all connected?

   18.505 feet

4. Amanda lives on a farm out in the country. It takes her 2.25 hours to drive to town and back. She usually goes to town twice a week to get supplies. How much time does Amanda spend driving if she makes 8 trips to town each month?

   18 hours

5. Destiny's favorite apples are $2.50 per pound at the grocery store. She bought 3.5 pounds of the apples. How much did she spend?

   $8.75

6. Jarvis bought a van that holds 21.75 gallons of gas and gets an average of 12.5 miles per gallon. How many miles can he expect to travel on a full tank?

   271.875 miles

7. Virginia has 12.25 cups of cupcake batter. If each cupcake uses 0.75 cups of batter, how many full-sized cupcakes can she make? How much batter will she have left over?

   16 cupcakes; $\frac{1}{3}$ cup left over

8. Grace worked 38.5 hours last week and earned $325.67. How much did she make per hour? Round your answer to the nearest cent.

   $8.46 per hour

Name____________________ (7.NS.A.2d)

Converting Fractions and Decimals

$\frac{1}{5} \longrightarrow 5\overline{)1.00} \longrightarrow \frac{1}{5} = 0.2$ Terminating

$\frac{1}{3} \longrightarrow 3\overline{)1.00} \longrightarrow \frac{1}{3} = 0.\overline{3}$ Repeating

Write each fraction as a decimal. Draw a line above repeating numbers in decimals.

1. $\frac{2}{3}$ **$0.\overline{6}$**
2. $\frac{1}{2}$ **0.5**
3. $\frac{4}{33}$ **$0.\overline{12}$**
4. $\frac{13}{15}$ **$0.8\overline{6}$**
5. $\frac{28}{35}$ **0.8**
6. $\frac{6}{15}$ **0.4**

Terminating Decimals

$0.50 = \frac{5}{100} = \frac{1}{2}$

Repeating Decimals

$x = 0.\overline{6}$ ← 1 digit repeats

$x = \frac{6}{9}$

$x = .\overline{23}$ ← 2 digits repeat

$x = \frac{23}{99}$

Write each decimal as a fraction. Write the answer in simplest form.

7. 0.6875 **$\frac{11}{16}$**
8. $0.\overline{35}$ **$\frac{35}{99}$**
9. 0.212 **$\frac{53}{250}$**
10. $0.\overline{48}$ **$\frac{16}{33}$**
11. 0.625 **$\frac{5}{8}$**
12. 0.54 **$\frac{27}{50}$**

Name____________________ (7.NS.A.2d)

Converting Fractions and Decimals

Write each fraction as a decimal. Draw a line above repeating numbers in decimals.

1. $\frac{9}{36}$ **0.25**
2. $\frac{8}{15}$ **$0.5\overline{3}$**
3. $\frac{30}{45}$ **$0.\overline{6}$**
4. $\frac{19}{57}$ **$0.\overline{3}$**
5. $\frac{45}{72}$ **0.625**
6. $\frac{21}{36}$ **$0.58\overline{3}$**
7. $\frac{10}{60}$ **$0.1\overline{6}$**
8. $\frac{32}{36}$ **$0.\overline{8}$**
9. $\frac{56}{63}$ **$0.\overline{8}$**

Write each decimal as a fraction. Write the answer in simplest form.

10. 0.345 **$\frac{69}{200}$**
11. 0.12 **$\frac{3}{25}$**
12. 0.555 **$\frac{111}{200}$**
13. 0.300 **$\frac{3}{10}$**
14. 0.942 **$\frac{471}{500}$**
15. 0.24 **$\frac{6}{25}$**
16. $0.34\overline{58}$ **$\frac{856}{2475}$**
17. 0.28 **$\frac{7}{25}$**
18. $0.13\overline{4}$ **$\frac{121}{900}$**

Answer Key

Name________________ (7.NS.A.2d)

Converting Fractions and Decimals

Write each fraction as a decimal. Draw a line above repeating numbers in decimals.

1. $\frac{5}{10}$ **0.5**
2. $\frac{12}{18}$ **$0.\overline{6}$**
3. $\frac{8}{24}$ **$0.\overline{3}$**
4. $\frac{56}{64}$ **0.875**
5. $\frac{48}{74}$ **$0.\overline{648}$**
6. $\frac{6}{22}$ **$0.\overline{27}$**
7. $\frac{16}{72}$ **$0.\overline{2}$**
8. $\frac{35}{55}$ **$0.\overline{63}$**
9. $\frac{6}{40}$ **0.15**
10. $\frac{4}{36}$ **$0.\overline{1}$**
11. $\frac{13}{39}$ **$0.\overline{3}$**
12. $\frac{18}{40}$ **0.45**

Write each decimal as a fraction. Write the answer in simplest form.

13. 0.34 $\frac{17}{50}$
14. 0.342 $\frac{171}{500}$
15. 0.708 $\frac{177}{250}$
16. 0.144 $\frac{18}{125}$
17. 0.65 $\frac{13}{20}$
18. 0.920 $\frac{23}{25}$
19. $0.33\overline{4}$ $\frac{301}{900}$
20. 0.438 $\frac{219}{500}$
21. 0.378 $\frac{189}{500}$
22. 0.82 $\frac{41}{50}$

© Carson-Dellosa • CD-104631 25

Name________________ (6.RP.A.1, 6.RP.A.2)

Ratios and Rates

A **ratio** is a statement of how two numbers compare. Ratios are used to compare the size of numbers or the rate of something. A **unit rate** is a ratio with a denominator of 1.

Ratios: 3 to 12 → $\frac{3}{12} = \frac{1}{4}$; 25:30 → $\frac{25}{30} = \frac{5}{6}$

Rate: $\frac{12 \text{ cookies}}{\$6}$ → $\frac{2 \text{ cookies}}{\$1}$

Write each ratio as a fraction. Write the answer in simplest form.

1. 66 to 40 $\frac{33}{20}$
2. 130 to 112 $\frac{65}{56}$
3. 110:112 $\frac{55}{56}$
4. 65:35 $\frac{13}{7}$
5. 21 to 84 $\frac{1}{4}$
6. 66:166 $\frac{33}{83}$
7. 30 to 323 $\frac{30}{323}$
8. 197 to 17 $\frac{197}{17}$

Find the unit rate. Round to the nearest hundredth.

9. 294 miles on 10 gallons — **29.4 miles / 1 gallon**
10. $0.84 for 12 ounces — **$0.07 / 1 ounce**
11. 4.5 inches of rain in 2 months — **2.25 inches / 1 month**
12. 400 meters in 24.5 seconds — **16.33 meters / 1 second**
13. $15 for a half dozen roses — **$2.50 / 1 rose**
14. $80 for 8 movie tickets — **$10 / 1 movie ticket**
15. 10 limes for $2 — **5 limes / $1**
16. $2.99 for 6 cans of lemonade — **$0.50 / 1 can**

26 © Carson-Dellosa • CD-104631

Name________________ (6.RP.A.1, 6.RP.A.2)

Ratios and Rates

A **ratio** shows how two numbers compare. It can use the word *to*, a colon (:), or look like a fraction. A **unit rate** is a ratio with a denominator of 1.

Write each ratio as a fraction. Write the answer in simplest form.

1. 20 to 70 $\frac{2}{7}$
2. 14 to 43 $\frac{14}{43}$
3. 121:108 $\frac{121}{108}$
4. 51:102 $\frac{1}{2}$
5. 34 out of 82 $\frac{17}{41}$
6. 112:224 $\frac{1}{2}$
7. 21 to 45 $\frac{7}{15}$
8. 40: 231 $\frac{40}{231}$

Find the unit rate. Round to the nearest hundredth.

9. 20 people in 4 cars — **5 people / 1 car**
10. 186 miles in 8 hours — **23.25 miles / 1 hour**
11. 82 hours for 18 projects — **4.56 hours / 1 project**
12. $950 earned in 5 weeks — **$190 / 1 week**
13. $2.32 for 16 ounces — **$0.15 / 1 ounce**
14. 578 miles in 9 hours — **64.22 miles / 1 hour**
15. 1200 cars in 5 dealerships — **240 cars / 1 dealership**
16. 132 miles on 4 gallons — **33 miles / 1 gallon**

© Carson-Dellosa • CD-104631 27

Name________________ (6.RP.A.1, 6.RP.A.2)

Ratios and Rates

Write each ratio as a fraction. Write the answer in simplest form.

1. $\frac{4}{6}$ $\frac{2}{3}$
2. $\frac{28}{36}$ $\frac{7}{9}$
3. 18 out of 30 $\frac{3}{5}$
4. 14 out of 56 $\frac{1}{4}$
5. 2:10 $\frac{1}{5}$
6. 14:21 $\frac{2}{3}$
7. $\frac{32}{64}$ $\frac{1}{2}$
8. 55:33 $\frac{5}{3}$
9. 25 out of 65 $\frac{5}{13}$
10. 100 out of 170 $\frac{10}{17}$

Find the unit rate. Round to the nearest hundredth.

11. 210.8 miles on 12.4 gallons — **17 miles / 1 gallon**
12. 2.5 inches of rain in 10 hours — **0.25 inches / 1 hour**
13. $10.20 for 15 pounds — **$0.68 / 1 pound**
14. $62.50 for 25 tickets — **$2.50 / 1 ticket**
15. $1.50 for 25 sticks of gum — **$0.06 / 1 stick**
16. $3.00 for 5 muffins — **$0.60 / 1 muffin**
17. $4.49 for 16 ounces — **$0.28 / 1 ounce**
18. 240 miles on 8 gallons — **30 miles / 1 gallon**
19. $425 for 5 days of work — **$85 / 1 day**
20. 90 miles in 2 hours — **45 miles / 1 hour**

28 © Carson-Dellosa • CD-104631

Answer Key

Name ______________________ (7.RP.A.2a, 7.RP.A.2c)

Proportions

A **proportion** is an equation that names 2 equivalent ratios.
Use cross-products to solve proportions.

$$\frac{n}{3} = \frac{6}{9}$$
$$n \times 9 = 3 \times 6$$
$$\frac{9n}{9} = \frac{18}{9}$$
$$n = 2$$

Solve each proportion. Use cross-products.

1. $\frac{3}{2} = \frac{n}{42}$ **n = 63**
2. $\frac{2}{5} = \frac{14}{x}$ **x = 35**
3. $\frac{n}{16} = \frac{5}{8}$ **n = 10**
4. $\frac{7}{5} = \frac{n}{20}$ **n = 28**
5. $\frac{3}{4} = \frac{x}{16}$ **x = 12**
6. $\frac{46}{92} = \frac{n}{100}$ **n = 50**
7. $\frac{2}{3} = \frac{24}{n}$ **n = 36**
8. $\frac{n}{2} = \frac{56}{112}$ **n = 1**
9. $\frac{9}{27} = \frac{x}{9.6}$ **x = 3.2**
10. $\frac{25}{5} = \frac{150}{x}$ **x = 30**

Name ______________________ (7.RP.A.2a, 7.RP.A.2c)

Proportions

$$\frac{2}{6} = \frac{x}{18}$$
$$2 \cdot 18 = 6x$$
$$\frac{36}{6} = \frac{6x}{6}$$
$$6 = x$$

Solve each proportion. Use cross-products.

1. $\frac{1}{4} = \frac{x}{8}$ **x = 2**
2. $\frac{20}{30} = \frac{5}{n}$ **n = 7.5**
3. $\frac{18}{24} = \frac{12}{x}$ **x = 16**
4. $\frac{80}{n} = \frac{48}{20}$ **n = $33.\overline{3}$**
5. $\frac{5}{5} = \frac{5x}{5}$ **x = 1**
6. $\frac{15}{45} = \frac{3}{x}$ **x = 9**
7. $\frac{1.8}{n} = \frac{3.6}{2.8}$ **n = 1.4**
8. $\frac{8}{n} = \frac{5}{2}$ **n = 3.2**
9. $\frac{8}{6} = \frac{n}{27}$ **n = 36**
10. $\frac{144}{6} = \frac{6n}{6}$ **n = 24**
11. $\frac{x}{3} = \frac{8}{8}$ **x = 3**
12. $\frac{36}{12} = \frac{n}{6}$ **n = 18**
13. $\frac{0.14}{0.07} = \frac{n}{1.5}$ **n = 3**
14. $\frac{5}{n} = \frac{6}{4}$ **n = $3.\overline{3}$**
15. $\frac{4}{5} = \frac{x}{5}$ **x = 4**
16. $\frac{16}{48} = \frac{x}{50}$ **x = $16.\overline{6}$**

Name ______________________ (7.RP.A.2a, 7.RP.A.2c)

Proportions

Solve each proportion. Use cross-products.

1. $\frac{182}{x} = \frac{91}{70}$ **x = 140**
2. $\frac{n}{20} = \frac{73}{10}$ **n = 146**
3. $\frac{186}{84} = \frac{93}{b}$ **b = 42**
4. $\frac{84}{136} = \frac{21}{a}$ **a = 34**
5. $\frac{c}{190} = \frac{29}{95}$ **c = 58**
6. $\frac{93}{x} = \frac{31}{65}$ **x = 195**
7. $\frac{a}{134} = \frac{30}{67}$ **a = 60**
8. $\frac{122}{a} = \frac{61}{100}$ **a = 200**
9. $\frac{x}{124} = \frac{92}{62}$ **x = 184**
10. $\frac{80}{188} = \frac{c}{94}$ **c = 40**
11. $\frac{12}{55} = \frac{b}{165}$ **b = 36**
12. $\frac{c}{46} = \frac{164}{92}$ **c = 82**
13. $\frac{b}{1} = \frac{15}{5}$ **b = 3**
14. $\frac{87}{29} = \frac{174}{x}$ **x = 58**
15. $\frac{94}{18} = \frac{x}{36}$ **x = 188**
16. $\frac{d}{35} = \frac{3}{105}$ **d = 1**
17. $\frac{86}{73} = \frac{x}{146}$ **x = 172**
18. $\frac{97}{95} = \frac{a}{190}$ **a = 194**
19. $\frac{99}{43} = \frac{198}{d}$ **d = 86**
20. $\frac{65}{b} = \frac{195}{111}$ **b = 37**

Name ______________________ (6.RP.A.3, 7.RP.A.1)

Problem Solving with Proportions

To solve problems with proportions, be sure to have the same unit in both numerators and the same unit in both denominators.

If a 9-pound turkey takes 180 minutes to cook, how long would a 6-pound turkey take to cook?

$$\frac{\text{pounds}}{\text{time}} = \frac{9}{180} = \frac{6}{m}$$
$$9m = 180 \cdot 6$$
$$\frac{9m}{9} = \frac{1080}{9} \qquad m = 120 \text{ minutes}$$

Solve each problem. Round each answer to the nearest cent or hundredth.

1. There are 220 calories in 4 ounces of beef. How many calories are there in 5 ounces?

   275 calories

2. If John can buy 8 liters of soft drinks at the store for $6.40, how much does it cost him to buy 12 liters?

   $9.60

3. Sherri bought a pack of pens that contained 15 pens. How many packs should she buy if she needs 240 pens?

   16 packs

4. Steve won his election by a margin of 7 to 2. His opponent had 3,492 votes. How many votes did Steve have?

   12,222 votes

5. A car traveled 325 miles in 5 hours. How far did the car travel in 9 hours?

   585 miles

6. A recipe asks for $1\frac{1}{2}$ cups of chocolate chips for 60 cookies. How many cups would be needed for 36 cookies?

   0.9 or $\frac{9}{10}$ cups

Answer Key

Name________________ (6.RP.A.3, 7.RP.A.1)

Problem Solving with Proportions

If 3 liters of juice cost \$3.75, how much does 9 liters cost?

$$\frac{\text{liters}}{\text{cost}} = \frac{3}{3.75} = \frac{9}{x}$$

$$3x = 3.75 \bullet 9$$

$$\frac{3x}{3} = \frac{33.75}{3} \qquad x = 11.25$$

9 liters cost \$11.25

Solve each problem. Round each answer to the nearest hundredth.

1. If 3 square feet of fabric cost \$3.75, what would 7 square feet cost?
   \$8.75
2. A 12-ounce bottle of soap costs \$2.50. How many ounces would be in a bottle that costs \$3.75?
   18 ounces
3. Four pounds of apples cost \$5.00. How much would 10 pounds of apples cost?
   \$12.50
4. A 12-ounce can of lemonade costs \$1.32. How much would a 16-ounce can of lemonade cost?
   \$1.76
5. J & S Jewelry company bought 800 bracelets for \$450.00. How much did each bracelet cost?
   \$0.56
6. A dozen peaches costs \$3.60. How much did each peach cost?
   \$0.30
7. A 32-pound box of cantaloupe costs \$24.40. How much would a 12-pound box cost?
   \$9.15
8. If a 10-pound turkey costs \$20.42, how much would a 21-pound turkey cost?
   \$42.88

Name________________ (6.RP.A.3, 7.RP.A.1)

Problem Solving with Proportions

Solve each problem. Round each answer to the nearest hundredth.

1. Buying 19 pounds of apples costs \$209. How much would 49 pounds cost?
   \$539
2. A car can travel 273 miles on 7.8 gallons of gasoline. How far can it travel on 10.4 gallons?
   364 miles
3. A boat can travel 83.82 miles on 27.94 gallons of gasoline. How much gasoline will it need to go 398.1 miles?
   132.7 gallons
4. A boat travels 381 miles in 27.2 hours. How far can it travel in 11.4 hours?
   159.68 miles
5. Buying 8 pounds of apples costs \$48.72. How much would 20 pounds cost?
   \$121.80
6. Buying 36 pounds of bananas costs \$136.80. How much would 1 pound cost?
   \$3.80
7. A boat travels 319 miles in 64 hours. How far can it travel in 77 hours?
   383.80 miles
8. A boat travels 432.08 miles in 31 hours. How much time will it take for the boat to travel 480.26 miles?
   34.46 hours
9. A car can travel 210 miles on 10 gallons of gasoline. How far can it travel on 18 gallons?
   378 miles
10. A boat travels 458.4 miles in 13.9 hours. How much time will it take for the boat to travel 493.6 miles?
    14.97 hours

Name________________ (7.NS.A.2d)

Percents

A percent is a ratio that compares a number to 100.

Fraction to Percent

$$\frac{1}{2} = \frac{x}{100}$$

$$2x = 100$$

$$x = 50$$

$$\frac{1}{2} = 50\%$$

Percent to Fraction

$$12\% = \frac{12}{100}$$

$$\frac{12}{100} = \frac{3}{25}$$

Decimal to Percent

$0.425 = 42.5\%$

When converting a decimal to a percent, multiply by 100, or move the decimal two places to the right.

Percent to Decimal

$68\% = 0.68$

When converting a percent to a decimal, divide by 100, or move the decimal two places to the left.

Write each fraction or decimal as a percent. Round each answer to the nearest hundredth.

1. $\frac{7}{21}$ **33.33%**
2. 10.8 **1080%**
3. 12.392 **1239.2%**
4. 523.32 **52332%**

Write each percent as a decimal and a fraction. Write fractions in simplest form.

5. 40% **0.4 and $\frac{2}{5}$**
6. 30.5% **0.305 and $\frac{61}{200}$**
7. 73% **0.73 and $\frac{73}{100}$**
8. 8.66% **0.0866 and $\frac{433}{5000}$**

Name________________ (7.NS.A.2d)

Percents

$$60\% = \frac{60}{100} = \frac{6}{10} = \frac{3}{5}$$

$$51.5\% = \frac{51.5}{100} = \frac{515}{1{,}000} = \frac{103}{200}$$

Write each percent as a fraction in simplest form. Write each fraction as a percent rounded to the nearest hundredth.

1. 0.68% **$\frac{17}{25}$**
2. 33.8% **$\frac{169}{500}$**
3. 8.6% **$\frac{43}{500}$**
4. 21.98% **$\frac{1{,}099}{5{,}000}$**
5. 4.934% **$\frac{2{,}467}{50{,}000}$**
6. $7\frac{4}{21}$ **719.05%**
7. 5.75% **$\frac{23}{400}$**
8. $3\frac{34}{88}$ **338.64%**
9. $4\frac{5}{46}$ **410.87%**
10. $2\frac{5}{7}$ **271.43%**
11. 364.69% **$\frac{36{,}469}{10{,}000}$**
12. $21\frac{7}{32}$ **2,121.88%**
13. 12.4% **$\frac{31}{250}$**
14. $6\frac{1}{2}$ **650%**
15. $1\frac{3}{4}$ **175%**
16. 2.98% **$\frac{149}{5{,}000}$**

Answer Key

Name ______________________ (7.NS.A.2d)

Percents

Write each fraction or decimal as a percent rounded to the nearest hundredth.

1. 2.3839 **238.39%**
2. $\frac{12}{19}$ **63.16%**
3. $\frac{5}{46}$ **10.87%**
4. $\frac{11}{23}$ **47.82%**
5. $\frac{4}{13}$ **30.77%**
6. 2.32 **232%**
7. 17.45 **1,745%**
8. 5.293 **529.3%**
9. 0.625 **62.5%**
10. $\frac{3}{2}$ **150%**

Complete the table by naming the fraction and decimal for each percent. Write fractions in simplest form.

	Percent	Fraction	Decimal
11.	25%	**1/4**	**0.25**
12.	5%	**1/20**	**0.05**
13.	28.5%	**57/200**	**0.285**
14.	8.75%	**7/80**	**0.0875**
15.	150%	**3/2**	**1.5**
16.	60%	**3/5**	**0.6**
17.	62.5%	**5/8**	**0.625**
18.	48%	**12/25**	**0.48**
19.	0.3%	**3/1000**	**0.003**
20.	0.6%	**3/500**	**0.006**

Name ______________________ (7.RP.A.3)

Problem Solving with Percents

50% of 60 = ____
$\frac{50}{100} = \frac{x}{60}$
100x = 3000
x = 30

____% of 40 = 20
$\frac{x}{100} = \frac{20}{40}$
40x = 2000
x = 50

40% of ____ = 30
$\frac{40}{100} = \frac{30}{x}$
40x = 3000
x = 75

Solve each problem using proportions.

1. 20% of 15 = **3**
2. 30% of 60 = **18**
3. 15% of 75 = **11.25**
4. 37% of 65 = **24.05**
5. 44% of 40 = **17.6**
6. **42.86%** % of 35 = 15
7. **28.57%** % of 70 = 20
8. **$16.\overline{6}$%** % of 48 = 8
9. 50% of **65** = 32.5
10. 45% of **9** = 4.05

Name ______________________ (7.RP.A.3)

Problem Solving with Percents

To solve word problems with percents, decide what you need to find the percent of by reading the problem. Then, use proportions to solve.

Each item at a sale was reduced by 25%. What was the regular price of a shirt that is reduced $9?

25% of regular price = $9

$\frac{25}{100} = \frac{9}{r}$

$\frac{25r}{25} = \frac{900}{25}$

$r = \$36$

Solve each problem using proportions.

1. 12% of **$633.\overline{3}$** = 76
2. 65% of **123.08** = 80
3. 20% of **375** = 75
4. 45% of **$333.\overline{3}$** = 150
5. 22% of **$154.\overline{54}$** = 34
6. 50% of **104** = 52
7. Molly made $30 in tips from her customers. If the total of her customers' bills was $200, what percent did her customers tip?
   15%
8. How many problems did Robert get right out of 40 if he received an 87.5% on his test?
   35
9. Mary has sold 90 boxes of cookies. If her goal was to sell 120 boxes, what percentage of her goal has she sold?
   75%
10. How much did Tom pay in income tax on a gross income of $50,000 if 9% of his income was taxed?
    $4,500

Name ______________________ (7.RP.A.3)

Problem Solving with Percents

Solve each problem using proportions.

1. **50%** % of 90 = 45
2. **19%** % of 100 = 19
3. 75% of 60 = **45**
4. 35% of **20** = 7
5. How much is a $48 shirt that is on sale for 25% off?
   $36
6. Janet borrowed $5,500 from the bank at an interest rate of 7.5% for 1 year. Assuming she pays it back on time, how much interest will she pay?
   $412.50
7. If 14 out of 24 students in a class are boys, what percent of the class are girls?
   $41.\overline{6}$%
8. At Pinetop Middle School, 372 students ride the bus to school. If this number is 60% of school enrollment, then how many students are enrolled at the school?
   620 students
9. If you purchase a game console that costs $625, how much sales tax will you pay if the rate is 7.375%?
   $46.09
10. A volleyball team won 75% of 80 games played. How many games did they win?
    60 games

Answer Key

Name____________________ (6.NS.C.5, 6.NS.C.6, 6.NS.C.7)

Integers and Absolute Value

Integers are positive and negative values. On a number line, positive integers are whole numbers to the right of 0, and negative integers are whole numbers to the left of 0.

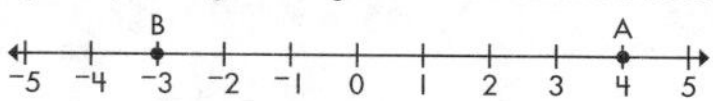

The coordinate of A on this number line is 4 and the coordinate of B on this number line is -3.

The **absolute value** of a number is the distance the number is from 0. It is shown by putting vertical lines on either side of an integer like this: |-7| The absolute value of this integer is 7 since it is 7 spaces from 0. Absolute value is always represented by a positive number.

Name the coordinate of each point on the number line using integers.

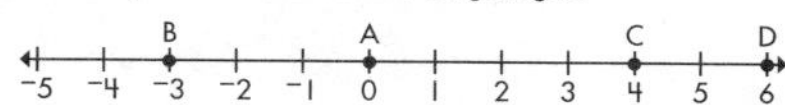

1. A **0**
2. B **-3**
3. C **4**
4. D **6**

Graph each integer on the number line below.

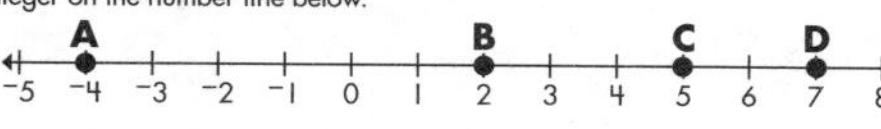

5. A = -4
6. B = 2
7. C = 5
8. D = 7

Write the absolute value of each number.

9. |-12| **12**
10. |0| **0**
11. |-10| **10**
12. |-15| **15**

Name____________________ (6.NS.C.5, 6.NS.C.6, 6.NS.C.7)

Integers and Absolute Value

Remember, positive integers are whole numbers to the right of 0 (point B), and negative integers are whole numbers to the left of 0 (point A).

When an integer is shown with 2 negative signs, like -(-2), it represents a positive value.

If absolute value is shown with a negative sign outside of the absolute value, -|4|, then the absolute value is -4.

Name the coordinate of each point on the number line using integers.

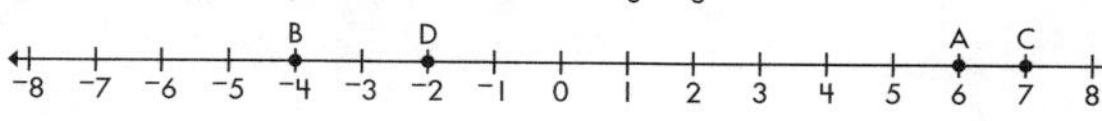

1. A **6**
2. B **-4**
3. C **7**
4. D **-2**

Graph each integer on the number line below.

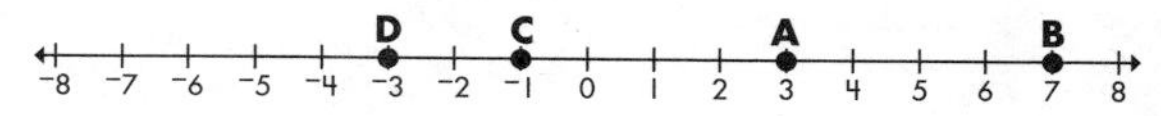

5. A = -(-3)
6. B = 7
7. C = -1
8. D = -(3)

Write the absolute value of each number.

9. -|-15| **-15**
10. |5| **5**
11. |-23| **23**
12. |12| **12**

Name____________________ (6.NS.C.5, 6.NS.C.6, 6.NS.C.7)

Integers and Absolute Value

Name the coordinate of each point on the number line using integers.

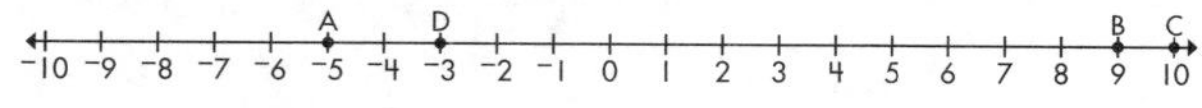

1. A **-5**
2. B **9**
3. C **10**
4. D **-3**

Graph each integer on the number line below.

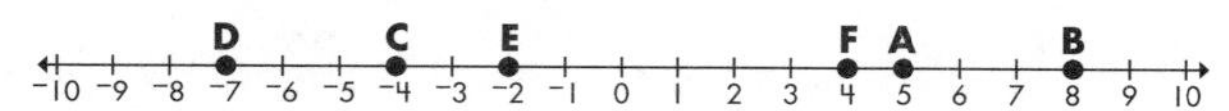

5. A = -(-5)
6. B = 8
7. C = -4
8. D = -(7)
9. E = -(2)
10. F = -(-4)

Write the absolute value of each number.

11. -|38| **-38**
12. -|-132| **-132**
13. |-96| **96**
14. |45| **45**
15. -|-73| **-73**
16. -|18| **-18**

Name____________________ (6.NS.C.7a)

Comparing and Ordering Integers

Integers can be compared based on their placement on a number line.

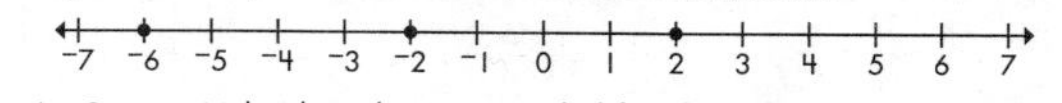

-6 < 2 -6 is less than 2 because it is to the left on the number line.

2 > -6 2 is greater than -6 because it is to the right on the number line.

Numbers can also be ordered based on where they fall on a number line.

Use <, >, or = to compare the numbers.

1. 10 **(>)** -12
2. |-3| **(=)** |3|
3. -11 **(<)** -2
4. 0 **(<)** |-2|
5. -5 **(<)** 0
6. -8 **(>)** -14

Order the numbers from least to greatest.

7. 1, -4, -2 **-4, -2, 1**
8. 76, -72, 45 **-72, 45, 76**
9. 3, -3, 4 **-3, 3, 4**
10. -6, 6, -7 **-7, -6, 6**

Answer Key

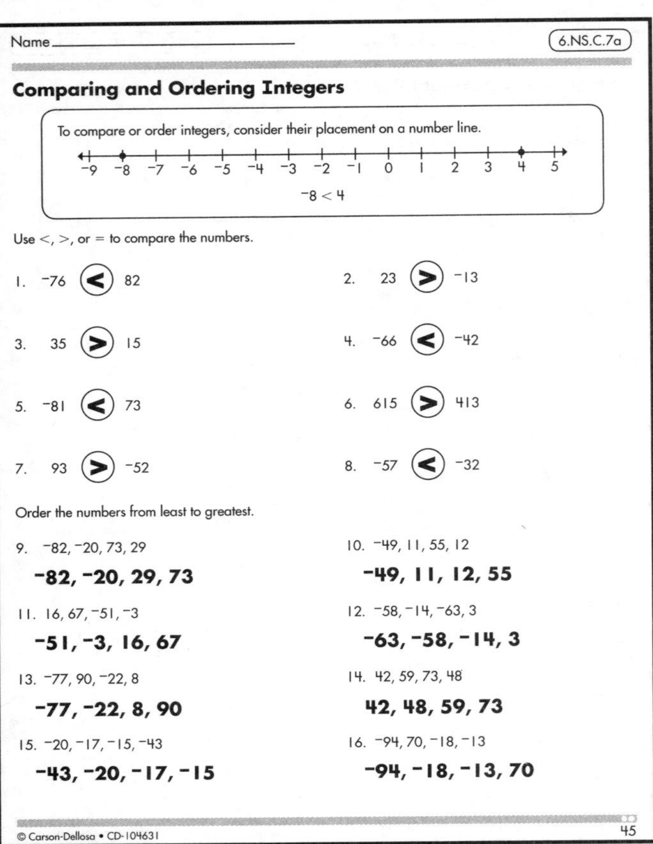

Name ____________________ (6.NS.C.7a)

Comparing and Ordering Integers

To compare or order integers, consider their placement on a number line.

−9 −8 −7 −6 −5 −4 −3 −2 −1 0 1 2 3 4 5

−8 < 4

Use <, >, or = to compare the numbers.

1. −76 **<** 82
2. 23 **>** −13
3. 35 **>** 15
4. −66 **<** −42
5. −81 **<** 73
6. 615 **>** 413
7. 93 **>** −52
8. −57 **<** −32

Order the numbers from least to greatest.

9. −82, −20, 73, 29 **−82, −20, 29, 73**
10. −49, 11, 55, 12 **−49, 11, 12, 55**
11. 16, 67, −51, −3 **−51, −3, 16, 67**
12. −58, −14, −63, 3 **−63, −58, −14, 3**
13. −77, 90, −22, 8 **−77, −22, 8, 90**
14. 42, 59, 73, 48 **42, 48, 59, 73**
15. −20, −17, −15, −43 **−43, −20, −17, −15**
16. −94, 70, −18, −13 **−94, −18, −13, 70**

© Carson-Dellosa • CD-104631 45

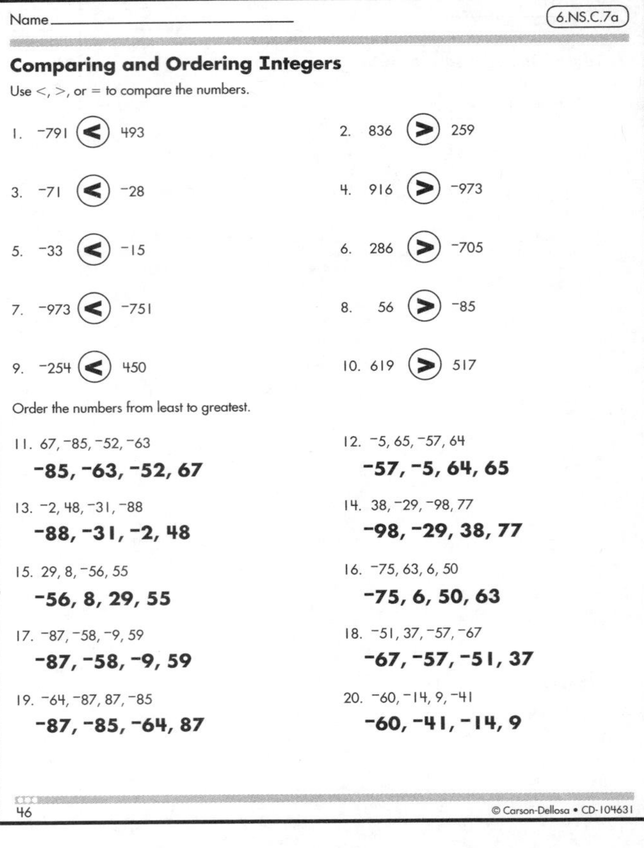

Name ____________________ (6.NS.C.7a)

Comparing and Ordering Integers

Use <, >, or = to compare the numbers.

1. −791 **<** 493
2. 836 **>** 259
3. −71 **<** −28
4. 916 **>** −973
5. −33 **<** −15
6. 286 **>** −705
7. −973 **<** −751
8. 56 **>** −85
9. −254 **<** 450
10. 619 **>** 517

Order the numbers from least to greatest.

11. 67, −85, −52, −63 **−85, −63, −52, 67**
12. −5, 65, −57, 64 **−57, −5, 64, 65**
13. −2, 48, −31, −88 **−88, −31, −2, 48**
14. 38, −29, −98, 77 **−98, −29, 38, 77**
15. 29, 8, −56, 55 **−56, 8, 29, 55**
16. −75, 63, 6, 50 **−75, 6, 50, 63**
17. −87, −58, −9, 59 **−87, −58, −9, 59**
18. −51, 37, −57, −67 **−67, −57, −51, 37**
19. −64, −87, 87, −85 **−87, −85, −64, 87**
20. −60, −14, 9, −41 **−60, −41, −14, 9**

46 © Carson-Dellosa • CD-104631

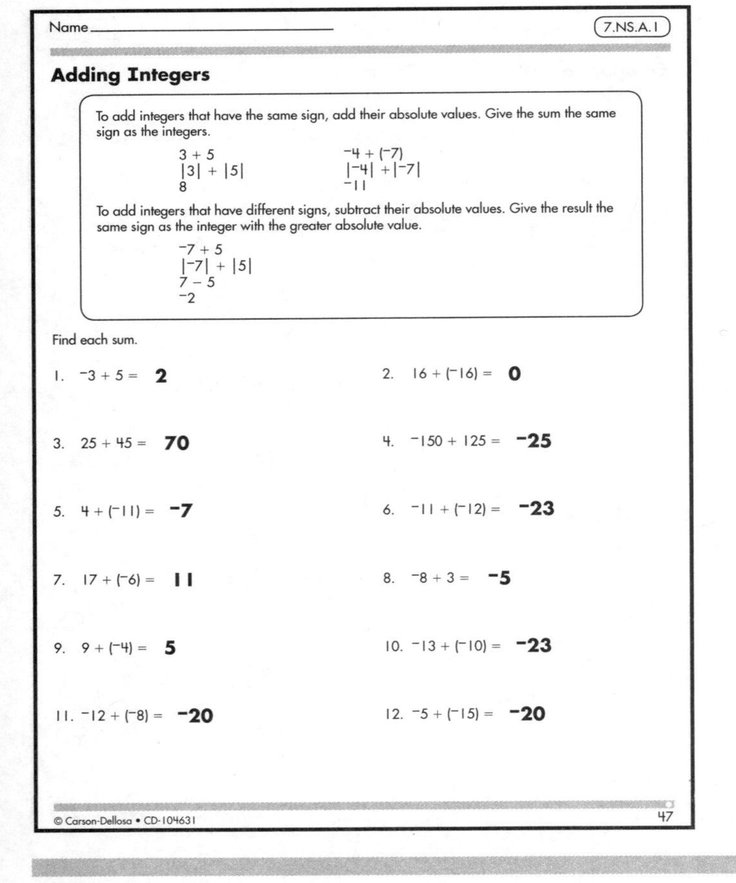

Name ____________________ (7.NS.A.1)

Adding Integers

To add integers that have the same sign, add their absolute values. Give the sum the same sign as the integers.

3 + 5
|3| + |5|
8

−4 + (−7)
|−4| + |−7|
−11

To add integers that have different signs, subtract their absolute values. Give the result the same sign as the integer with the greater absolute value.

−7 + 5
|−7| + |5|
7 − 5
−2

Find each sum.

1. −3 + 5 = **2**
2. 16 + (−16) = **0**
3. 25 + 45 = **70**
4. −150 + 125 = **−25**
5. 4 + (−11) = **−7**
6. −11 + (−12) = **−23**
7. 17 + (−6) = **11**
8. −8 + 3 = **−5**
9. 9 + (−4) = **5**
10. −13 + (−10) = **−23**
11. −12 + (−8) = **−20**
12. −5 + (−15) = **−20**

© Carson-Dellosa • CD-104631 47

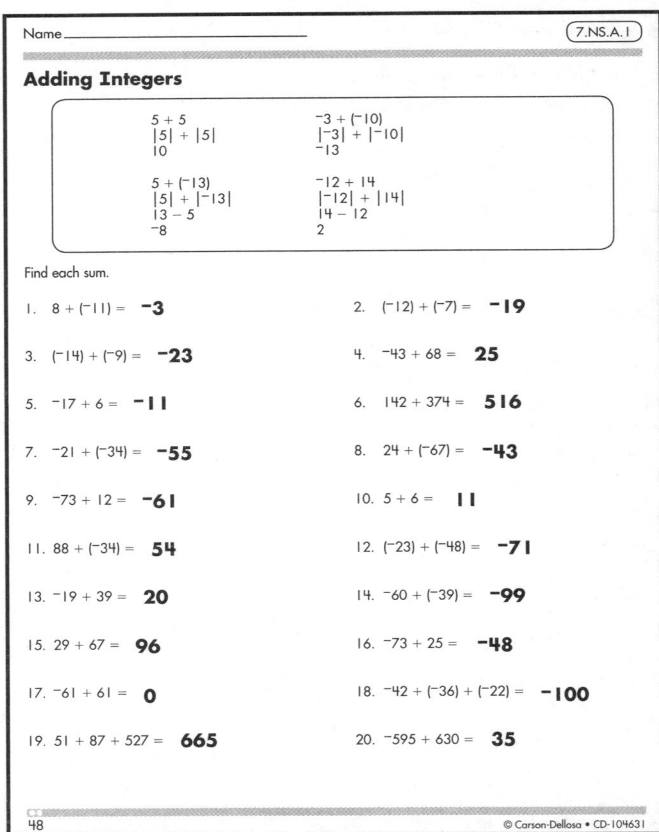

Name ____________________ (7.NS.A.1)

Adding Integers

5 + 5
|5| + |5|
10

−3 + (−10)
|−3| + |−10|
−13

5 + (−13)
|5| + |−13|
13 − 5
−8

−12 + 14
|−12| + |14|
14 − 12
2

Find each sum.

1. 8 + (−11) = **−3**
2. (−12) + (−7) = **−19**
3. (−14) + (−9) = **−23**
4. −43 + 68 = **25**
5. −17 + 6 = **−11**
6. 142 + 374 = **516**
7. −21 + (−34) = **−55**
8. 24 + (−67) = **−43**
9. −73 + 12 = **−61**
10. 5 + 6 = **11**
11. 88 + (−34) = **54**
12. (−23) + (−48) = **−71**
13. −19 + 39 = **20**
14. −60 + (−39) = **−99**
15. 29 + 67 = **96**
16. −73 + 25 = **−48**
17. −61 + 61 = **0**
18. −42 + (−36) + (−22) = **−100**
19. 51 + 87 + 527 = **665**
20. −595 + 630 = **35**

48 © Carson-Dellosa • CD-104631

Answer Key

Name ____________________ (7.NS.A.1)

Adding Integers

Find each sum.

1. −6,607 + 4,362 = **−2,245**
2. (−21) + (−59) + (−828) = **−908**
3. 23 + 54 + 56 = **133**
4. 17,985 + (−22,581) = **−4,596**
5. 67,888 + (−78,952) = **−11,064**
6. 213 + 375 = **588**
7. (−16) + (−16) + (−16) = **−48**
8. 48 + (−56) = **−8**
9. −99 + 94 = **−5**
10. (−13) + (−65) + (−78) + (−332) = **−488**
11. 12 + 45 + 332 = **389**
12. 45,908 + (−12,921) = **32,987**
13. −112,956 + 564,258 = **451,302**
14. (−20) + (−68) = **−88**
15. 15 + 41 + 7 = **63**
16. −34,132 + 81,323 = **47,191**
17. 850 + (−828) = **22**
18. −563,937 + 76,412 = **−487,525**
19. (−14) + (−34) + (−67) = **−115**
20. −34 + (−76) = **−110**

 49

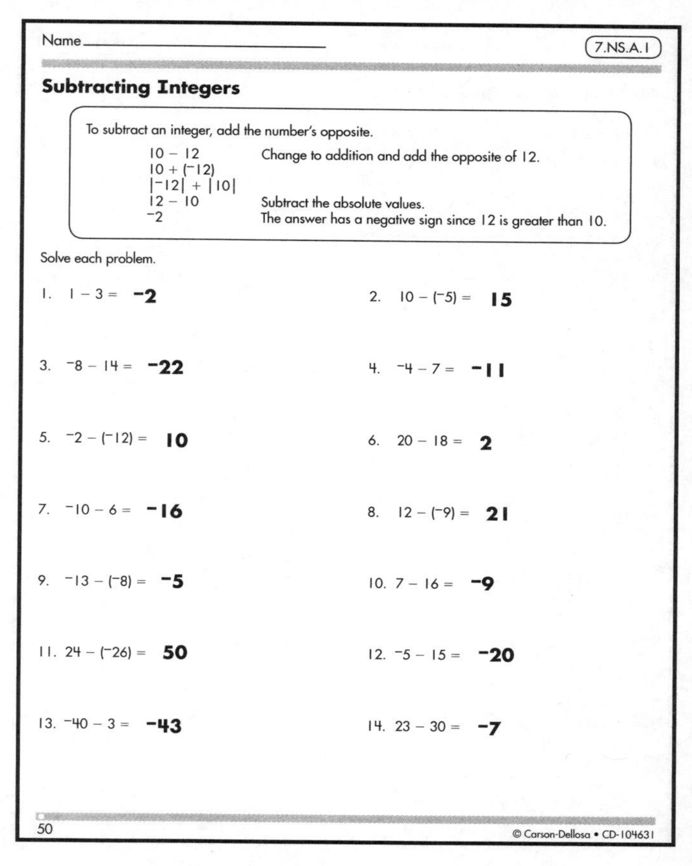
Name ____________________ (7.NS.A.1)

Subtracting Integers

To subtract an integer, add the number's opposite.

10 − 12 10 + (−12) |−12| + |10|	Change to addition and add the opposite of 12.
12 − 10 −2	Subtract the absolute values. The answer has a negative sign since 12 is greater than 10.

Solve each problem.

1. 1 − 3 = **−2**
2. 10 − (−5) = **15**
3. −8 − 14 = **−22**
4. −4 − 7 = **−11**
5. −2 − (−12) = **10**
6. 20 − 18 = **2**
7. −10 − 6 = **−16**
8. 12 − (−9) = **21**
9. −13 − (−8) = **−5**
10. 7 − 16 = **−9**
11. 24 − (−26) = **50**
12. −5 − 15 = **−20**
13. −40 − 3 = **−43**
14. 23 − 30 = **−7**

50

Name ____________________ (7.NS.A.1)

Subtracting Integers

6 − 10 = 6 + (−10) = −4 (Add the opposite)

6 − (−10) = 6 + 10 = 16 (Add the opposite)

Solve each problem.

1. −8 − 3 = **−11**
2. 56 − (−65) = **121**
3. −52 − (−34) = **−18**
4. −19 − (−13) = **−6**
5. 42 − 23 = **19**
6. 77 − 22 = **55**
7. 17 − 26 = **−9**
8. −594 − (−73) = **−521**
9. −117 − 29 = **−146**
10. −749 − 629 = **−1,378**
11. 19 − (−342) = **361**
12. 2,567 − (−492) = **3,059**
13. 5,762 − 2,144 = **3,618**
14. 121 − 154 = **−33**
15. −8 − (−27) = **19**
16. −87 − 129 = **−216**
17. 45 − 75 = **−30**
18. 688 − 456 = **232**
19. 187 − (−48) = **235**
20. 157 − (−452) = **609**

 51

Name ____________________ (7.NS.A.1)

Subtracting Integers

Solve each problem.

1. 7 − 15 = **−8**
2. 319 − (−749) = **1,068**
3. −18 − 6 = **−24**
4. −60 − 17 = **−77**
5. −45 − (−45) = **0**
6. −54 − (−95) = **41**
7. −154 − 56 = **−210**
8. −625 − 127 = **−752**
9. −21 − (−45) = **24**
10. 564 − (−373) = **937**
11. 3 − (−67) = **70**
12. 6,593 − (−4,132) = **10,725**
13. 0 − 15 = **−15**
14. 108,762 − (−95,671) = **204,433**
15. −3 − (−7) = **4**
16. −9,774 − 8,834 = **−18,608**
17. −5 − (−67) = **62**
18. −23 − 56 = **−79**
19. −44 − 57 = **−101**
20. −475,824 − (−153,198) = **−322,626**
21. −36 − 69 = **−105**
22. −934 − (−672) = **−262**
23. −630,805 − (−512,156) = **−118,649**
24. 899,342 − (−392,231) = **1,291,573**

52

Answer Key

Name ____________________ (7.NS.A.2a, 7.NS.A.2c)

Multiplying Integers

To multiply integers, simply multiply the two numbers. The product of two integers with the same sign is positive and the product of two integers with different signs is negative.

$12 \times 2 = 24$, $^-4 \times (^-9) = 36$ } The product is positive since both integers have the same sign.

$^-5 \times 7 = {}^-35$ The product is negative since the integers have different signs.

Solve each problem.

1. $7 \times 4 =$ **28**
2. $^-3 \times (^-5) =$ **15**
3. $^-7 \times 0 =$ **0**
4. $6 \times (^-8) =$ **⁻48**
5. $^-9 \times 9 =$ **⁻81**
6. $^-2 \times (^-11) =$ **22**
7. $^-10 \times 7 =$ **⁻70**
8. $^-16 \times (^-2) =$ **32**
9. $^-12 \times (^-11) =$ **132**
10. $^-15 \times 0 =$ **0**
11. $5 \times 8 =$ **40**
12. $8 \times (^-10) =$ **⁻80**
13. $6 \times (^-20) =$ **⁻120**
14. $^-3 \times 13 =$ **⁻39**

© Carson-Dellosa • CD-104631 53

Name ____________________ (7.NS.A.2a, 7.NS.A.2c)

Multiplying Integers

$(3)(3) = 9$ $(^-2)(^-3) = 6$ like signs = positive

$(^-3)(3) = {}^-9$ $(^-2)(3) = {}^-6$ unlike signs = negative

When multiplying more than 2 integers, an even number of negative signs will give a positive answer. An odd number of negative signs will give a negative answer.

Solve each problem.

1. $(33)(^-123)(12) =$ **⁻48,708**
2. $(^-434)(^-7) =$ **3,038**
3. $(15)(^-4) =$ **⁻60**
4. $(^-5)(^-28)(^-23) =$ **⁻3,220**
5. $(30)(5) =$ **150**
6. $(13)(^-28) =$ **⁻364**
7. $(^-72)(43) =$ **⁻3,096**
8. $(^-3)(9) =$ **⁻27**
9. $(56)(12) =$ **672**
10. $(14)(^-33)(2) =$ **⁻924**
11. $(32)(^-48) =$ **⁻1,536**
12. $(20)(^-3)(23)(^-3) =$ **4,140**
13. $(^-39)(^-58) =$ **2,262**
14. $(12)(^-12)(2)(^-33) =$ **9,504**
15. $(^-20)(^-10)(2)(3) =$ **1,200**
16. $(37)(^-90) =$ **⁻3,330**
17. $(121)(^-10)(21) =$ **⁻25,410**
18. $(^-9)(^-88)(^-7) =$ **⁻5,544**
19. $(^-13)(^-13) =$ **169**
20. $(^-32)(^-22)(^-45) =$ **⁻31,680**

54 © Carson-Dellosa • CD-104631

Name ____________________ (7.NS.A.2a, 7.NS.A.2c)

Multiplying Integers

Solve each problem.

1. $(8)(^-99)(^-22)(^-7) =$ **⁻121,968**
2. $(^-9)(^-82)(^-7) =$ **⁻5,166**
3. $(^-7)(3) =$ **⁻21**
4. $(^-6)(^-9) =$ **54**
5. $(^-10)(^-5)(^-3) =$ **⁻150**
6. $(5)(^-4)(^-3) =$ **60**
7. $(^-7)(^-2)(^-5) =$ **⁻70**
8. $(^-7)(^-14)(144) =$ **14,112**
9. $(^-17)(^-2) =$ **34**
10. $(^-2)(^-13)(^-4) =$ **⁻104**
11. $(^-8)(^-9) =$ **72**
12. $(^-1)(22)(^-33)(44) =$ **31,944**
13. $(21)(^-22) =$ **⁻462**
14. $(^-85)(^-215) =$ **18,275**
15. $(4)(111)(^-1) =$ **⁻444**
16. $(213)(4)(18) =$ **15,336**
17. $(^-19)(^-38) =$ **722**
18. $(^-5)(^-100)(^-302) =$ **⁻151,000**
19. $(1)(^-41)(^-6) =$ **246**
20. $(^-33)(213) =$ **⁻7,029**

© Carson-Dellosa • CD-104631 55

Name ____________________ (7.NS.A.2b, 7.NS.A.2c)

Dividing Integers

To divide integers, simply divide the two numbers. The quotient of two integers with the same sign is positive and the quotient of two integers with different signs is negative.

$^-10 \div (^-5) = 2$, $60 \div 5 = 12$ } The quotient is positive since both integers have the same sign.

$^-45 \div 9 = {}^-5$ The quotient is negative since the integers have different signs.

Solve each problem.

1. $9 \div 3 =$ **3**
2. $45 \div (^-15) =$ **⁻3**
3. $^-28 \div 4 =$ **⁻7**
4. $0 \div (^-7) =$ **0**
5. $^-16 \div (^-8) =$ **2**
6. $^-26 \div (^-13) =$ **2**
7. $^-121 \div (^-11) =$ **11**
8. $0 \div 15 =$ **0**
9. $100 \div (^-20) =$ **⁻5**
10. $20 \div (^-5) =$ **⁻4**
11. $(^-9) \div 3 =$ **⁻3**
12. $^-38 \div (^-19) =$ **2**
13. $96 \div 16 =$ **6**
14. $^-21 \div (^-7) =$ **3**

56 © Carson-Dellosa • CD-104631

Answer Key

Name________________ (7.NS.A.2b, 7.NS.A.2c)

Dividing Integers

$\frac{-18}{-3} = 6$ like signs = positive

$24 \div (-4) = -6$ unlike signs = negative

Solve each problem.

1. $100 \div (-4) =$ **−25**
2. $\frac{-18}{18} =$ **−1**
3. $-60 \div 3 =$ **−20**
4. $\frac{-104}{8} =$ **−13**
5. $120 \div (-6) =$ **−20**
6. $\frac{-77}{7} =$ **−11**
7. $88 \div (-22) =$ **−4**
8. $\frac{36}{-9} =$ **−4**
9. $-188 \div 4 =$ **−47**
10. $\frac{168}{21} =$ **8**
11. $144 \div (-12) =$ **−12**
12. $\frac{-50}{-5} =$ **10**
13. $80 \div (-5) =$ **−16**
14. $-36 \div 6 =$ **−6**
15. $72 \div 4 =$ **18**
16. $\frac{169}{-13} =$ **−13**
17. $\frac{210}{-10} =$ **−21**
18. $\frac{-20}{4} =$ **−5**
19. $-150 \div 6 =$ **−25**
20. $\frac{-288}{-12} =$ **24**

Name________________ (7.NS.A.2b, 7.NS.A.2c)

Dividing Integers

Solve each problem.

1. $-14 \div 14 =$ **−1**
2. $\frac{-77}{11} =$ **−7**
3. $60 \div (-10) =$ **−6**
4. $-160 \div (-40) =$ **4**
5. $-72 \div 9 =$ **−8**
6. $\frac{-80}{10} =$ **−8**
7. $\frac{-755}{-5} =$ **151**
8. $\frac{-72}{8} =$ **−9**
9. $-54 \div (-9) =$ **6**
10. $\frac{-35}{-7} =$ **5**
11. $-195 \div (-65) =$ **3**
12. $\frac{-468}{26} =$ **−18**
13. $-150 \div (-50) =$ **3**
14. $\frac{-253}{11} =$ **−23**
15. $189 \div (-21) =$ **−9**
16. $\frac{66}{-2} =$ **−33**
17. $75 \div (-3) =$ **−25**
18. $\frac{-84}{-7} =$ **12**
19. $-210 \div (-5) =$ **42**
20. $\frac{-552}{-23} =$ **24**
21. $-94 \div 2 =$ **−47**
22. $\frac{-310}{5} =$ **−62**
23. $-125 \div 5 =$ **−25**
24. $\frac{-258}{-3} =$ **86**

Name________________ (7.EE.A.1)

Properties of Operations

The **commutative property** states that the order of the operation can be changed and the result is still the same. Addition and multiplication both have the commutative property.

$2 + 3 = 3 + 2$ $4 \times 2 = 2 \times 4$

The **associative property** states that a set of numbers can be grouped in different ways without changing the result. Addition and multiplication both have the associative property.

$(3 + 5) + 7 = 3 + (5 + 7)$ $(12 \times 13) \times 14 = 12 \times (13 \times 14)$

The **distributive property** allows for multiples before operations to be distributed and for factors within multiples to be taken out.

$3(7 + 2) = (3)(7) + (3)(2)$
$21 + 6 = 27$

$3x + x = 24$
$x(3 + 1) = 24$

Rewrite each expression using the commutative property.

1. $6 + y = 12$ **$y + 6 = 12$**
2. $a + 17 = 21$ **$17 + a = 21$**
3. $x + 4 = 12$ **$4 + x = 12$**

Rewrite each expression using the associative property.

4. $8 + (2 + 17)$ **$(8 + 2) + 17$**
5. $(3 \times -4) \times 250$ **$3 \times (-4 \times 250)$**
6. $68 + (32 + 54)$ **$(68 + 32) + 54$**

Rewrite each expression using the distributive property. Do not simplify.

7. $4(12 + 15)$ **$(4)(12) + (4)(15)$**
8. $(10 + 13)z$ **$10z + 13z$**
9. $7r + 8r + 2$ **$r(7 + 8) + 2$**

Name________________ (7.EE.A.1)

Properties of Operations

The **commutative property** states that the order of the operation can be changed and the result is still the same.

The **associative property** states that a set of numbers can be grouped in different ways without changing the result.

The **distributive property** allows for multiples before operations to be distributed and for factors within multiples to be taken out.

Rewrite each expression using the commutative property.

1. $h \times 15 = 75$ **$15 \times h = 75$**
2. $56 = y \times 7$ **$56 = 7 \times y$**
3. $11 + k = 32$ **$k + 11 = 32$**
4. $3 = r + 8$ **$3 = 8 + r$**

Rewrite each expression using the associative property. Then, solve the expression.

5. $(3 \times 25) \times 4$ **$3 \times (25 \times 4) = 300$**
6. $(21 + 45) + 55$ **$21 + (45 + 55) = 121$**
7. $(12 \times 5) \times 20$ **$12 \times (5 \times 20) = 1,200$**
8. $75 + (25 + 19)$ **$(75 + 25) + 19 = 119$**

Rewrite each expression using the distributive property.

9. $3a + 6b$ **$3(a + 2b)$**
10. $x(6 + 8)$ **$6x + 8x$**
11. $2(5x + 8y)$ **$10x + 16y$**
12. $2(b + 4) + 8b$ **$2b + 8 + 8b$**

Answer Key

Name____________________ 7.EE.A.1

Properties of Operations

Rewrite each expression using the commutative property.

1. $39 = 13 \times f$ **$39 = f \times 13$**
2. $9 + y = 45$ **$y + 9 = 45$**
3. $3x + 4y = 18$ **$4y + 3x = 18$**
4. $8 \times 3b = 456$ **$3b \times 8 = 456$**
5. $9 \times 3 \times 6 = 162$ **Answers will vary but should include 3, 6, and 9 multiplied in any other order.**

Rewrite each expression using the associative property. Then, solve the expression.

6. $^-20 \times (50 \times {}^-29)$ **$(^-20 \times 50) \times {}^-29 = 29{,}000$**
7. $5 \times (10 \times 18)$ **$(5 \times 10) \times 18 = 900$**
8. $(400 + 60) + 98$ **$400 + (60 + 98) = 558$**
9. $(3 \times 20) \times 5$ **$3 \times (20 \times 5) = 300$**
10. $18 + (9 + 91)$ **$(18 + 9) + 91 = 118$**

Rewrite each expression using the distributive property. Simplify when possible.

11. $a(7 + 1) + 15$ **$7a + a + 15 = 8a + 15$**
12. $k + 5 + 7 + 3k$ **$4k + 12 = 4(k + 3)$**
13. $2c + 6c + 9(c + 3)$ **$8c + 9c + 27 = 17c + 27$**
14. $10(7 \times r)$ **$70 \times 10r$**
15. $(6y + 5y) + 1$ **$y(6 + 5) + 1$**

 61

Name____________________ 5.OA.A.1, 6.EE.A.2c

Order of Operations

When an expression contains more than one operation, it is important to use the order of operations to find its value.

1. Solve inside parentheses.
2. Multiply and divide from left to right.
3. Add and subtract from left to right.

$2(8 + 6) - 7 \times 3$
$2(14) - 7 \times 3$
$28 - 21$
7

Solve each problem using order of operations.

1. $(18 - 2) \div 4 =$ **4**
2. $12 - (4 + 7) =$ **1**
3. $7 \times (3 + 4) =$ **49**
4. $12 \div 3 \times 2 =$ **8**
5. $(11 + 4) \div 5 =$ **3**
6. $(10 \times 4) \div (2 \times 2) =$ **10**
7. $30 \div 6 - 1 =$ **4**
8. $42 \div (5 + 2) \times 3 =$ **18**
9. $56 \div (7 \times 2) =$ **4**
10. $(12 + 8) \div 4 =$ **5**

62

Name____________________ 5.OA.A.1, 6.EE.A.2c

Order of Operations

$^-4 \times 2 + 2 = {}^-8 + 2 = {}^-6$

$2\frac{1}{4} \div (4 + 8) = \frac{9}{8} \div 12 = \frac{9}{8} \times \frac{1}{12} = \frac{9}{96}$ or $\frac{3}{32}$

Solve each problem using order of operations.

1. $2 \times 3\,[7 + (6 \div 2)] =$ **60**
2. $\frac{2}{3}(-15 - 4) =$ **$-12\frac{2}{3}$**
3. $-8 \div (-2) + 5 \times (\frac{-1}{2}) - 25 \div 5 =$ **$-3\frac{1}{2}$**
4. $-30 \div 6 + 4\frac{1}{5} =$ **$-\frac{4}{5}$**
5. $(9\frac{1}{3} + 4\frac{1}{3}) \div 6 - (-12) =$ **$14\frac{5}{18}$**
6. $\frac{[(60 \div 4) + 35]}{(^-12 + 35)} =$ **$2\frac{4}{23}$**
7. $\frac{3}{4}[(-15 + 4) + (6 + 7) \div (-3)] =$ **$-11\frac{1}{2}$**
8. $3[-3(2 - 8) - 6] =$ **36**

 63

Name____________________ 5.OA.A.1, 6.EE.A.2c

Order of Operations

Solve each problem using order of operations.

1. $3 \times 3[2 - (9 \div 3)] =$ **$^-9$**
2. $\frac{1}{2}(-12 + 6) =$ **$^-3$**
3. $(5\frac{1}{5} - 6\frac{1}{5}) \times 6 - (-16) =$ **10**
4. $-20 \div 3 + 2\frac{2}{3} =$ **-4**
5. $-5 \div (-3) - 2 \times (-\frac{1}{3}) - 21 \div 7 =$ **$-\frac{2}{3}$**
6. $\frac{[(20 \div 2) + 10]}{(^-10 + 20 + 30)} =$ **$\frac{1}{2}$**
7. $2[-5(4 - 12) - 3] =$ **74**
8. $\frac{1}{2}[(-12 - 2) + (1 + 8) \div (-8)] =$ **$-7\frac{9}{16}$**
9. $[(3 \times 3) - (30 \div 6)] + (-27) - 13 =$ **-36**
10. $2 \div [(4 \div 2) + (-32 \div 8)] =$ **-1**
11. $30 \times [(3 \times 9) - (21 \div 7)] + (-32) =$ **688**

64

Answer Key

Name ____________________ 6.EE.A.2a

Using Variables

An expression that contains variables, numbers, and at least one operation is an **algebraic expression**. An algebraic expression can be evaluated by replacing the variables with an assigned value.

If $x = 2$ and $y = 5$:

$$6x - 2y$$
$$6(2) - 2(5)$$
$$12 - 10$$
$$2$$

Evaluate each expression if $x = 5$, $y = 2$, and $z = 8$.

1. $2z - 3x =$ **1**
2. $\frac{6x}{y} + z =$ **23**
3. $2z - xy =$ **6**
4. $10x - (4y + z) =$ **34**
5. $4x - (y + z) =$ **10**
6. $\frac{7z}{x} + y =$ **$13\frac{1}{5}$**
7. $6z + 7y - 3x =$ **47**
8. $2z + 3x + 4y =$ **39**

Name ____________________ 6.EE.A.2a

Using Variables

If $w = \frac{1}{5}$, $x = 4$, and $y = {}^-5$,

then $3x(5w + 2y) = 3 \cdot 4[5(\frac{1}{5}) + 2(^-5)] = 12[(1 + (^-10)] = 12(^-9) = {}^-108$

Evaluate each expression if $w = \frac{1}{5}$, $x = 4$, and $y = {}^-5$.

1. $y(w + 7) =$ **$^-36$**
2. $3w + 4(x - y) =$ **$36\frac{3}{5}$**
3. $6[w + (^-y)] =$ **$31\frac{1}{5}$**
4. $wx + x + 6xy =$ **$^-115\frac{1}{5}$**
5. $5(w - 2y) =$ **51**
6. $w(x + y) =$ **$^-\frac{1}{5}$**
7. $w(xw + xy) =$ **$^-3\frac{21}{25}$**
8. $7w - (xy + 3) =$ **$18\frac{2}{5}$**
9. $3w(3y + 5x) =$ **3**
10. $wx(3w + 3y - 6) =$ **$^-16\frac{8}{25}$**
11. $3w - 4x =$ **$^-15\frac{2}{5}$**
12. $10y(4y + 2w) =$ **980**
13. $8x + (^-12x) =$ **$^-16$**
14. $4w - 7x + 3y - 2w =$ **$^-42\frac{3}{5}$**

Name ____________________ 6.EE.A.2a

Using Variables

Evaluate each expression.

1. $5 + 6k + 2h =$ if $k = {}^-7$ and $h = 9$ — **$^-19$**
2. $\frac{z}{5} - 3f =$ if $z = 20$ and $f = {}^-6$ — **22**
3. $^-2(5r + 6b) =$ if $b = 6$ and $r = 4$ — **$^-112$**
4. $^-7(8 - 6z) + 5c =$ if $z = {}^-3$ and $c = 4$ — **$^-162$**
5. $3 + 8(^-2k - 7z) =$ if $z = {}^-8$ and $k = {}^-2$ — **483**
6. $9b + k =$ if $b = 7$ and $k = {}^-8$ — **55**
7. $\frac{^-4}{b} + 2x =$ if $b = 2$ and $x = 5$ — **8**
8. $5 - \frac{16}{s} - 3w =$ if $s = 4$ and $w = {}^-6$ — **19**
9. $n + 5x =$ if $n = 2$ and $x = {}^-5$ — **$^-23$**
10. $^-2(4z - 7b) =$ if $z = 4$ and $b = 3$ — **10**
11. $6w - 7(^-9 + 5h) =$ if $h = {}^-4$ and $w = 2$ — **215**
12. $^-9(^-8z - 2) + 6n =$ if $z = {}^-7$ and $n = 5$ — **$^-456$**
13. $^-5(9c + 6s) =$ if $s = {}^-8$ and $c = {}^-7$ — **555**
14. $\frac{18}{x} - 3 - 8r =$ if $x = 6$ and $r = 2$ — **$^-16$**
15. $^-7(6b - 3f) =$ if $b = 4$ and $f = {}^-2$ — **$^-210$**
16. $5s - 8k - 3 =$ if $s = {}^-7$ and $k = 4$ — **$^-70$**

Name ____________________ 7.EE.B.4a

Solving Addition Equations

To find a solution to an addition equation, the variable must be isolated. Subtract addends from both sides of the equation to isolate the variable. Then, solve.

$$r + 12 = 67$$
$$r + 12 - 12 = 67 - 12$$
$$r = 55$$

Solve each equation for the variable.

1. $x + 7 = 12$ **$x = 5$**
2. $24 = h + 3$ **$h = 21$**
3. $6 + a = 15$ **$a = 9$**
4. $21 + y = 15$ **$y = {}^-6$**
5. $n + 10 = 14$ **$n = 4$**
6. $64 + h = 36$ **$h = {}^-28$**
7. $^-13 = z + 7$ **$z = {}^-20$**
8. $4 + r = {}^-6$ **$r = {}^-10$**

Answer Key

Name ______________________ 7.EE.B.4a

Solving Addition Equations

$$1.4 = {}^{-}2.4 + x$$
$$1.4 + 2.4 = {}^{-}2.4 + 2.4 + x$$
$$3.8 = 0 + x$$
$$3.8 = x$$

Solve each equation for the variable.

1. $x + (^{-}5\frac{3}{4}) = {}^{-}10\frac{1}{4}$ **$x = {}^{-}4\frac{1}{2}$**
2. $^{-}35 = x + 35$ **$x = {}^{-}70$**
3. $w + 79 = {}^{-}95$ **$w = {}^{-}174$**
4. $-\frac{1}{4} + x = -\frac{1}{4}$ **$x = 0$**
5. $7 + c = {}^{-}14$ **$c = {}^{-}21$**
6. $^{-}4.5 = 9\frac{1}{2} + c$ **$c = {}^{-}14$**
7. $^{-}21 = y + 18$ **$y = {}^{-}39$**
8. $22 = c + (^{-}13)$ **$c = 35$**
9. $^{-}9 + r = 22$ **$r = 31$**
10. $x + (^{-}8) = 9$ **$x = 17$**
11. $3.5 = n + 4.6$ **$n = {}^{-}1.1$**
12. $^{-}2\frac{1}{2} + k = {}^{-}3\frac{5}{7}$ **$k = {}^{-}1\frac{3}{14}$**
13. $^{-}2{,}929 + y = 4{,}242$ **$y = 7{,}171$**
14. $z + 5.2 = 7.1$ **$z = 1.9$**

Name ______________________ 7.EE.B.4a

Solving Addition Equations

Solve each equation for the variable.

1. $-\frac{3}{4} + j = {}^{-}3\frac{1}{2}$ **$j = {}^{-}2\frac{3}{4}$**
2. $^{-}8\frac{3}{7} + y = {}^{-}9\frac{2}{5}$ **$y = -\frac{34}{35}$**
3. $^{-}3.1 = 4\frac{3}{4} + e$ **$e = {}^{-}7\frac{17}{20}$**
4. $b + (^{-}9) = 36$ **$b = 45$**
5. $5.77 = q + 9$ **$q = {}^{-}3.23$**
6. $^{-}3{,}282 = n + 1{,}111$ **$n = {}^{-}4{,}393$**
7. $d + 12 = {}^{-}18$ **$d = {}^{-}30$**
8. $f + (^{-}3\frac{1}{4}) = {}^{-}7\frac{1}{4}$ **$f = {}^{-}4$**
9. $x + 9 = {}^{-}22$ **$x = {}^{-}31$**
10. $18 + k = 16$ **$k = {}^{-}2$**
11. $2 + x = {}^{-}20$ **$x = {}^{-}22$**
12. $(^{-}24.64) + c = {}^{-}7.45$ **$c = 17.19$**
13. $(^{-}4.48) + f = 17.6$ **$f = 22.08$**
14. $(^{-}22) + r = {}^{-}17$ **$r = 5$**
15. $(^{-}1) + g = {}^{-}11$ **$g = {}^{-}10$**
16. $(^{-}17) + p = {}^{-}24$ **$p = {}^{-}7$**
17. $14 + x = 18$ **$x = 4$**
18. $-\frac{5}{2} + w = -\frac{9}{22}$ **$w = 2\frac{1}{11}$**

Name ______________________ 7.EE.B.4a

Solving Subtraction Equations

To find a solution to a subtraction equation, the variable must be isolated. Add one value to both sides of the equation to isolate the variable. Then, solve.

$$x - 16 = 32$$
$$x - 16 + 16 = 32 + 16$$
$$x = 48$$

Solve each equation for the variable.

1. $b - 14 = 51$ **$b = 65$**
2. $36 = d - 13$ **$d = 49$**
3. $x - 28 = 72$ **$x = 100$**
4. $11 = x - 1$ **$x = 12$**
5. $^{-}600 = c - (^{-}400)$ **$c = {}^{-}1000$**
6. $y - 8 = 8$ **$y = 16$**
7. $a - 15 = {}^{-}21$ **$a = {}^{-}6$**
8. $2 - y = {}^{-}4$ **$y = 6$**

Name ______________________ 7.EE.B.4a

Solving Subtraction Equations

$$32 = x - (^{-}8)$$
$$32 = x + 8$$
$$32 - 8 = x + 8 - 8$$
$$24 = x + 0$$
$$24 = x$$

Solve each equation for the variable.

1. $^{-}2{,}547 = n - 5{,}534$ **$n = 2{,}987$**
2. $^{-}44 = m - 32$ **$m = {}^{-}12$**
3. $d - (^{-}8) = 45$ **$d = 37$**
4. $-\frac{1}{3} - g = -\frac{1}{3}$ **$g = 0$**
5. $^{-}15 = p - 6$ **$p = {}^{-}9$**
6. $3.65 = n - 7$ **$n = 10.65$**
7. $34 = b - (^{-}2)$ **$b = 32$**
8. $a + (^{-}4\frac{1}{3}) = {}^{-}15\frac{1}{3}$ **$a = {}^{-}11$**
9. $f - 16 = {}^{-}32$ **$f = {}^{-}16$**
10. $x - 8 = 34$ **$x = 42$**
11. $^{-}3.4 = h - 8.5$ **$h = 5.1$**
12. $^{-}2.2 = 8\frac{4}{5} + d$ **$d = {}^{-}11$**
13. $^{-}3\frac{2}{3} + k = {}^{-}6\frac{3}{4}$ **$k = {}^{-}3\frac{1}{12}$**
14. $z - (^{-}21.5) = {}^{-}2.356$ **$z = {}^{-}23.856$**

Answer Key

Name________________________ 7.EE.B.4a

Solving Subtraction Equations

Solve each equation for the variable.

1. $11.4 - k = 5.2$ **$k = 6.2$**
2. $^{-}56 = c - (^{-}8)$ **$c = {}^{-}64$**
3. $^{-}6.7 = y - 27$ **$y = 20.3$**
4. $w - (^{-}43.7) = {}^{-}3.674$ **$w = {}^{-}47.374$**
5. $31 = d - 12$ **$d = 43$**
6. $y - 1 = \frac{5}{14}$ **$y = 1\frac{5}{14}$**
7. $z - 14 = {}^{-}24$ **$z = {}^{-}10$**
8. $s - (^{-}3) = {}^{-}22$ **$s = {}^{-}25$**
9. $v - (^{-}15.69) = 25.6$ **$v = 9.91$**
10. $k - \frac{15}{13} = \frac{6}{13}$ **$k = 1\frac{8}{13}$**
11. $p - 23 = {}^{-}13$ **$p = 10$**
12. $x - (^{-}4) = {}^{-}19$ **$x = {}^{-}23$**
13. $e - (^{-}14) = 21$ **$e = 7$**
14. $h - (^{-}13.94) = 21.06$ **$h = 7.12$**
15. $a - 2 = 10$ **$a = 12$**
16. $j - (^{-}25) = 1$ **$j = {}^{-}24$**
17. $d - (^{-}11.31) = 6.64$ **$d = {}^{-}4.67$**
18. $x - \frac{21}{22} = {}^{-}\frac{2}{11}$ **$x = \frac{17}{22}$**

Name________________________ 7.EE.B.4a

Solving Multiplication Equations

To find a solution to a multiplication equation, the variable must be isolated. Divide the known factor from both sides of the equation to isolate the variable. Then, solve.

$$4x = 84$$
$$4x \div 4 = 84 \div 4$$
$$x = 21$$

Solve each equation for the variable.

1. $9x = 63$ **$x = 7$**
2. $^{-}96b = 96$ **$b = {}^{-}1$**
3. $4d = {}^{-}36$ **$d = {}^{-}9$**
4. $^{-}64 = {}^{-}16y$ **$y = 4$**
5. $^{-}25x = {}^{-}125$ **$x = 5$**
6. $3x = 21$ **$x = 7$**
7. $6n = {}^{-}72$ **$n = {}^{-}12$**
8. $12a = 156$ **$a = 13$**
9. $^{-}12r = 12$ **$r = {}^{-}1$**
10. $54 = {}^{-}9e$ **$e = {}^{-}6$**

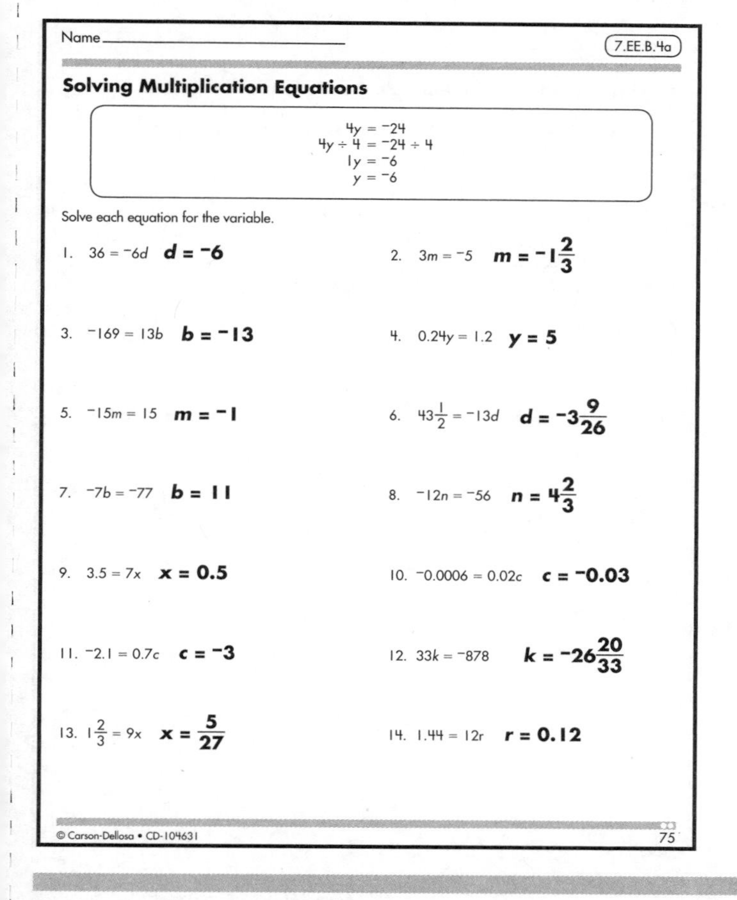

Name________________________ 7.EE.B.4a

Solving Multiplication Equations

$$4y = {}^{-}24$$
$$4y \div 4 = {}^{-}24 \div 4$$
$$1y = {}^{-}6$$
$$y = {}^{-}6$$

Solve each equation for the variable.

1. $36 = {}^{-}6d$ **$d = {}^{-}6$**
2. $3m = {}^{-}5$ **$m = {}^{-}1\frac{2}{3}$**
3. $^{-}169 = 13b$ **$b = {}^{-}13$**
4. $0.24y = 1.2$ **$y = 5$**
5. $^{-}15m = 15$ **$m = {}^{-}1$**
6. $43\frac{1}{2} = {}^{-}13d$ **$d = {}^{-}3\frac{9}{26}$**
7. $^{-}7b = {}^{-}77$ **$b = 11$**
8. $^{-}12n = {}^{-}56$ **$n = 4\frac{2}{3}$**
9. $3.5 = 7x$ **$x = 0.5$**
10. $^{-}0.0006 = 0.02c$ **$c = {}^{-}0.03$**
11. $^{-}2.1 = 0.7c$ **$c = {}^{-}3$**
12. $33k = {}^{-}878$ **$k = {}^{-}26\frac{20}{33}$**
13. $1\frac{2}{3} = 9x$ **$x = \frac{5}{27}$**
14. $1.44 = 12r$ **$r = 0.12$**

Name________________________ 7.EE.B.4a

Solving Multiplication Equations

Solve each equation for the variable.

1. $^{-}250 = 25s$ **$s = {}^{-}10$**
2. $18\frac{1}{3} = {}^{-}12w$ **$w = {}^{-}1\frac{19}{36}$**
3. $72 = {}^{-}8r$ **$r = {}^{-}9$**
4. $15n = {}^{-}3$ **$n = {}^{-}\frac{1}{5}$**
5. $^{-}12q = 12$ **$q = {}^{-}1$**
6. $^{-}0.0009 = 0.03q$ **$q = {}^{-}0.03$**
7. $56 = {}^{-}7e$ **$e = {}^{-}8$**
8. $43d = {}^{-}734$ **$d = {}^{-}17\frac{3}{43}$**
9. $^{-}4.5 = 9h$ **$h = {}^{-}0.5$**
10. $0.48y = 2.4$ **$y = 5$**
11. $^{-}6.7 = {}^{-}0.137k$ **$k = 48.905$**
12. $9h = {}^{-}90$ **$h = {}^{-}10$**
13. $2\frac{4}{6} = 5e$ **$e = \frac{8}{15}$**
14. $^{-}13g = {}^{-}78$ **$g = 6$**

Answer Key

Name ____________ 7.EE.B.4a

Solving Division Equations

To find a solution to a division equation, the variable must be isolated. Multiply each side of the equation by the divisor to isolate the variable. Then, solve.

$$f \div 21 = 2$$
$$f \div 21 \times 21 = 2 \times 21$$
$$f = 42$$

Solve each equation for the variable.

1. $\frac{a}{-6} = 2$ **$a = -12$**
2. $\frac{b}{40} = -3$ **$b = -120$**
3. $\frac{r}{15} = 20$ **$r = 300$**
4. $-77 = \frac{d}{11}$ **$d = -847$**
5. $\frac{f}{3.6} = 16$ **$f = 57.6$**
6. $-6 = \frac{x}{6}$ **$x = -36$**
7. $\frac{h}{9} = 63$ **$h = 567$**
8. $\frac{m}{5} = 22$ **$m = 110$**
9. $\frac{n}{2} = 8$ **$n = 16$**
10. $\frac{b}{12} = -12$ **$b = -144$**

Name ____________ 7.EE.B.4a

Solving Division Equations

$$\frac{2}{3}x = 6$$
$$3 \cdot \frac{2}{3}x = 6 \cdot 3$$
$$2x = 18$$
$$x = 9$$

Solve each equation for the variable.

1. $\frac{m}{7} = 42$ **$m = 294$**
2. $-12 = \frac{d}{4}$ **$d = -48$**
3. $0.9 = \frac{k}{81}$ **$z = 72.9$**
4. $(\frac{1}{7})n = -28$ **$n = -196$**
5. $(\frac{3}{4})z = 144$ **$z = 192$**
6. $\frac{r}{17} = -23$ **$r = -391$**
7. $\frac{e}{4} = -36$ **$e = -144$**
8. $-3 = (\frac{1}{3})x$ **$x = -9$**
9. $-15 = \frac{x}{3}$ **$x = -45$**
10. $\frac{x}{7} = 56$ **$x = 392$**
11. $(\frac{1}{12})c = 0.6$ **$c = 7.2$**
12. $(\frac{3}{7})h = 4.5$ **$h = 10.5$**
13. $\frac{x}{4.1} = 18$ **$x = 73.8$**
14. $(\frac{1}{4})c = -8$ **$c = -32$**

Name ____________ 7.EE.B.4a

Solving Division Equations

Solve each equation for the variable.

1. $\frac{f}{4.44} = -3.63$ **$f = -16.1172$**
2. $\frac{e}{-5} = 5$ **$e = -25$**
3. $\frac{h}{-2} = -10$ **$h = 20$**
4. $\frac{q}{2.45} = 10.67$ **$q = 26.1415$**
5. $\frac{s}{-1.55} = 10.53$ **$s = -16.3215$**
6. $\frac{e}{(\frac{-8}{13})} = \frac{23}{8}$ **$e = -\frac{23}{13}$**
7. $\frac{k}{5} = -4$ **$k = -20$**
8. $\frac{z}{10.81} = 2.39$ **$z = 25.8359$**
9. $\frac{a}{-2.25} = -6.16$ **$a = 13.86$**
10. $\frac{e}{(\frac{-1}{25})} = \frac{25}{19}$ **$x = -\frac{1}{19}$**
11. $\frac{y}{2.86} = 12.22$ **$y = 34.9492$**
12. $\frac{b}{-10} = 2$ **$b = -20$**
13. $\frac{z}{13} = -1$ **$z = -13$**
14. $\frac{m}{-2} = 10$ **$m = -20$**

Name ____________ 7.EE.B.4a

Solving Equations with Multiple Operations

To solve a problem with multiple operations, the variable must be isolated. Isolate the variable by using addition, subtraction, multiplication, and division.

$$2y - 10 = 30$$
$$2y - 10 + 10 = 30 + 10$$
$$\frac{2y}{2} = \frac{40}{2}$$
$$y = 20$$

Solve each equation for the variable. Write the answer in simplest form.

1. $14 = 6c - 4$ **$c = 3$**
2. $13n - 13 = -12$ **$n = \frac{1}{13}$**
3. $5x - 5 = -10$ **$x = -1$**
4. $23x - 12 = -33$ **$x = -\frac{21}{23}$**
5. $10 = 3y + 5$ **$y = 1\frac{2}{3}$**
6. $-42 = 6b + 8$ **$b = -8\frac{1}{3}$**
7. $-23 = 3e - (-9)$ **$e = -10\frac{2}{3}$**
8. $16 + 4y = -32$ **$y = -12$**
9. $-8r - 9 = -24$ **$r = 1\frac{7}{8}$**
10. $16 + \frac{r}{2} = -11$ **$r = -54$**

Answer Key

Name__________________ (7.EE.B.4a)

Solving Equations with Multiple Operations

$$2x + 5 = 3x + 6$$
$$2x - 3x + 5 = 3x - 3x + 6$$
$$-x + 5 = 6$$
$$-x + 5 - 5 = 6 - 5$$
$$-x = 1$$
$$\frac{-x}{-1} = \frac{1}{-1}$$
$$x = -1$$

Solve each equation for the variable. Write the answer in simplest form.

1. $4n - 6 = 6n + 14$ **$n = -10$**
2. $-6g + 12 = 2g + 12$ **$g = 0$**
3. $-d + 9 = d + 5$ **$d = 2$**
4. $23b + 9 = 4b + 66$ **$b = 3$**
5. $7y - 7 = 5y + 13$ **$y = 10$**
6. $-8z = 27 + z$ **$z = -3$**
7. $10w + 6 = 6w - 15$ **$w = -5\frac{1}{4}$**
8. $13y - 26 = 7y + 22$ **$y = 8$**
9. $-r - 5 = 1 - 2r$ **$r = 6$**
10. $3m - 8 = 5m + 8$ **$m = -8$**
11. $4x - 7 = 2x + 7$ **$x = 7$**
12. $2e + 20 = 4e - 12$ **$e = 16$**
13. $18 + 2p = 8p - 13$ **$p = 5\frac{1}{6}$**
14. $4h + 10 = 2h - 22$ **$h = -16$**

Name__________________ (7.EE.B.4a)

Solving Equations with Multiple Operations

Solve each equation for the variable. Write the answer in simplest form.

1. $-2j + 6 = 2j - 8$ **$j = 3\frac{1}{2}$**
2. $4h + 6 = 2h - 8$ **$h = -7$**
3. $9w + 9 = 3w - 15$ **$w = -4$**
4. $45 = -9(e + 8)$ **$e = -13$**
5. $35 = -7d$ **$d = -5$**
6. $12k + 13 = 8k + 33$ **$k = 5$**
7. $6u = 28 - z$ **$z = 4$**
8. $7(9 - 6j) = -63$ **$j = 3$**
9. $4(y - 8) = -12$ **$y = 5$**
10. $-6(36 - 10b) + 8 = 32$ **$b = 4$**
11. $-14k = 56$ **$k = -4$**
12. $9(8c - 9) = -351$ **$c = -3\frac{3}{4}$**
13. $11g = 121$ **$g = 11$**
14. $\frac{m}{2.5} = 22$ **$m = 55$**
15. $24 = 4[(\frac{h}{2}) - 7]$ **$h = 26$**
16. $-6 = \frac{b}{6}$ **$b = -36$**
17. $4g + 12 = 6g - 4$ **$g = 8$**
18. $(\frac{2}{5})h = -20$ **$h = -50$**

Name__________________ (5.OA.A.2, 7.EE.B.3, 7.EE.B.4a)

Writing Algebraic Expressions

Use the words in word problems to help you understand which operations should be used.

Three times a number increased by 5	$3x + 5$
A number increased by 3	$x + 3$
A number divided by 2	$x \div 2$ or $\frac{x}{2}$
The product of 2 and 6	2×6

Write the algebraic expression.

1. Two-fifths of a number decreased by 3 **$(\frac{2}{5})x - 3$**
2. Twelve times a number decreased by 4 **$12x - 4$**
3. Eight times the difference between x and 7 **$8(x - 7)$**
4. The product of 3 and a number increased by 6 **$3x + 6$**
5. One-third times a number increased by 7 **$\frac{1}{3}x + 7$**
6. Four increased by 7 times a number **$4 + 7x$**
7. Seven times the sum of twice a number and 16 **$7(2x + 16)$**
8. Eleven times the sum of a number and 5 times the number **$11(x + 5x)$**
9. Five times a number plus 6 times the number **$5x + 6x$**
10. The quotient of a number and 5 decreased by 3 **$\frac{x}{5} - 3$**
11. Four times the sum of a number and 8 **$4(x + 8)$**
12. A number increased by 9 times the number **$x + 9x$**

Name__________________ (5.OA.A.2, 7.EE.B.3, 7.EE.B.4a)

Writing Algebraic Expressions

Five more than 2 times a number is 21. What is the number?

$$5 + 2x = 21$$
$$5 - 5 + 2x = 21 - 5$$
$$2x = 16$$
$$x = 8$$

Write an equation for each expression and solve.

1. Two times the sum of a number and 5 is 26. What is the number?
   $2(x + 5) = 26, x = 8$
2. The quotient of a number and 3 decreased by 6 is 7. What is the number?
   $\frac{x}{3} - 6 = 7, x = 39$
3. The product of a number and 4 increased by 7 is 5. What is the number?
   $4x + 7 = 5, x = -\frac{1}{2}$
4. Three more than 5 times a number is 58. What is the number?
   $3 + 5x = 58, x = 11$
5. Six more than a number is −31. What is the number?
   $6 + x = -31, x = -37$
6. Two-thirds of a number increased by 3 is 11. What is the number?
   $(\frac{2}{3})x + 3 = 11, x = 12$
7. Ten less than 2 times a number is 24. What is the number?
   $2x - 10 = 24, x = 17$

Answer Key

Name ______________________ (5.OA.A.2, 7.EE.B.3, 7.EE.B.4a)

Writing Algebraic Expressions

Write an equation for each expression and solve.

1. One half of a number is 14 more than 2 times the number. Find the number.

   $\frac{1}{2}x = 14 + 2x,\ x = -9\frac{1}{3}$

2. Forty increased by 4 times a number is 8 less than 6 times the number. Find the number.

   $40 + 4x = 6x - 8,\ x = 24$

3. Nineteen increased by 3 times a number is 4 less than 4 times the number. Find the number.

   $19 + 3x = 4x - 4,\ x = 23$

4. Four times the sum of a number and 3 is 7 times the number decreased by 3. Find the number.

   $4(x + 3) = 7x - 3,\ x = 5$

5. Twice a number decreased by 44 is 6 times the sum of the number and 3 times the number. Find the number.

   $2x - 44 = 6(x + 3x),\ x = -2$

6. Thirty decreased by 3 times a number is 6 less than 3 times the number. Find the number.

   $30 - 3x = 3x - 6,\ x = 6$

7. Twelve increased by 6 times a number is 6 less than 7 times the number. Find the number.

   $12 + 6x = 7x - 6,\ x = 18$

© Carson-Dellosa • CD-104631 85

Name ______________________ (6.EE.A.3)

Simplifying Expressions

Simplify expressions by using the distributive property first.
$2(x + 3y) = 2x + 2 \times 3y = 2x + 6y$
If possible, use addition and subtraction to combine like terms.
$4a - 3a + 7z = (4 - 3)a + 7z = a + 7z$

Expand each expression using the distributive property.

1. $4(2r + 6y) =$ $8r + 24y$
2. $2(3p - 3p) =$ $6p - 6p$
3. $-6(2b + 3c) =$ $-12b - 18c$
4. $7(-c + 6d) =$ $-7c + 42d$
5. $2(x - 12) =$ $2x - 24$
6. $12(2y + 5w) =$ $24y + 60w$

Combine like terms.

7. $3x + 3y + xy - 6xy + 5x + (-4y) =$ $8x - 5xy - y$
8. $12p + 4pd - 2p + 6pd =$ $10p + 10pd$
9. $2x + 3xy + 4x + 5xy + 6x =$ $12x + 8xy$
10. $4x - 2x + 6xy + 21x + (-9xy) - 9 =$ $23x - 3xy - 9$
11. $-3n + 12 - 4n =$ $-7n + 12$
12. $5e + 6ed + 5d - 7ed + 7 =$ $5e + 5d - ed + 7$

86 © Carson-Dellosa • CD-104631

Name ______________________ (6.EE.A.3)

Simplifying Expressions

Remember, simplify expressions by using the distributive property first. Then, combine like terms when possible.

Expand each expression using the distributive property.

1. $3(2 + r) =$ $6 + 3r$
2. $3(w - 4) =$ $3w - 12$
3. $8[y + (-2x)] =$ $8y - 16x$
4. $5(2 + 13y) =$ $10 + 65y$
5. $2k[-xy + (-8)] =$ $-2kxy - 16k$
6. $-7(2x + 9) =$ $-14x - 63$
7. $5(2y + 5x) =$ $10y + 25x$
8. $3(x + 2y + z) =$ $3x + 6y + 3z$

Combine like terms.

9. $3xy + 13xy - 12xy =$ $4xy$
10. $2r + 4ry - 5r + 3x - 4ry =$ $-3r + 3x$
11. $10ax - 2ax + 12x - 2a + (-2x) =$ $8ax + 10x - 2a$
12. $7a + a - 2a + 3ab - ab + 2ab =$ $6a + 4ab$
13. $7r + 2r - 4 =$ $9r - 4$
14. $5m + 2m + 40m + m + 17 =$ $48m + 17$
15. $23x - 7x + 4x =$ $20x$
16. $4x + 3y - 3xy + 6x - 2xy =$ $10x + 3y - 5xy$

© Carson-Dellosa • CD-104631 87

Name ______________________ (6.EE.A.3)

Simplifying Expressions

Simplify each expression by using the distributive property and combining like terms.

1. $7(2x + 5y) + 6xy - 6(3xy + 5x) =$ $-16x - 12xy + 35y$
2. $10p + 5pd - 2p + 6pd =$ $8p + 11pd$
3. $9x - 4x + 2x + 8(6x + 2x) =$ $71x$
4. $3(x - 5x) + 2(xy + 7x) + (-7xy) =$ $2x - 5xy$
5. $3c - 4bc + (-7b) + 3[2bc - (-b)] =$ $3c + 2bc - 4b$
6. $-2a - [-3(a + 7)] - 4(-a + b) =$ $5a + 21 - 4b$
7. $2t + 12t - 4(n + 4n) =$ $14t - 20n$
8. $6r + 5r - 8p + 6p + 7(2r - 4r) =$ $-3r - 2p$
9. $3n(x - y) + 3n(x + y) - 2 =$ $6nx - 2$
10. $3[h - (-k)] + 2[-3h + (-4k)] =$ $-3h - 5k$
11. $-2(g + 5g) + -2[6f - (-12g)] =$ $-12f - 36g$
12. $4(2x + 2y) - 2[3xy - (-5x)] =$ $-2x + 8y - 6xy$
13. $5m + 3mn - (-9n) + 2(m - n) =$ $3mn + 7m + 7n$
14. $5xy - 12xy + 12xy - 9(x + y) =$ $5xy - 9x - 9y$

88 © Carson-Dellosa • CD-104631

Answer Key

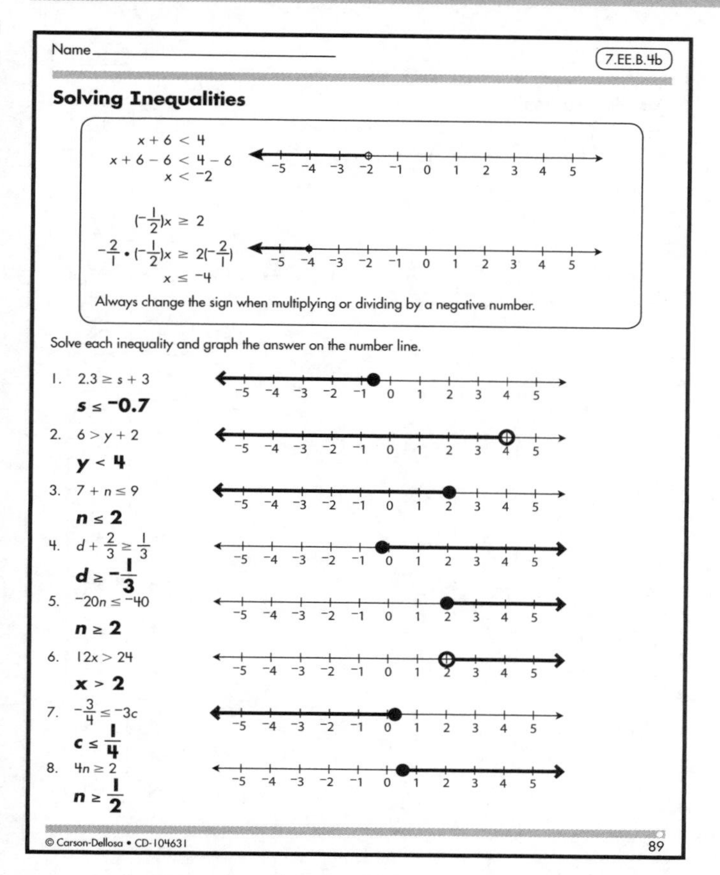

Name____________________ (7.EE.B.4b)

Solving Inequalities

$x + 6 < 4$
$x + 6 - 6 < 4 - 6$
$x < -2$

$(-\frac{1}{2})x \geq 2$
$-\frac{2}{1} \cdot (-\frac{1}{2})x \geq 2(-\frac{2}{1})$
$x \leq -4$

Always change the sign when multiplying or dividing by a negative number.

Solve each inequality and graph the answer on the number line.

1. $2.3 \geq s + 3$ — **$s \leq -0.7$**
2. $6 > y + 2$ — **$y < 4$**
3. $7 + n \leq 9$ — **$n \leq 2$**
4. $d + \frac{2}{3} \geq \frac{1}{3}$ — **$d \geq -\frac{1}{3}$**
5. $-20n \leq -40$ — **$n \geq 2$**
6. $12x > 24$ — **$x > 2$**
7. $-\frac{3}{4} \leq -3c$ — **$c \leq \frac{1}{4}$**
8. $4n \geq 2$ — **$n \geq \frac{1}{2}$**

© Carson-Dellosa • CD-104631 89

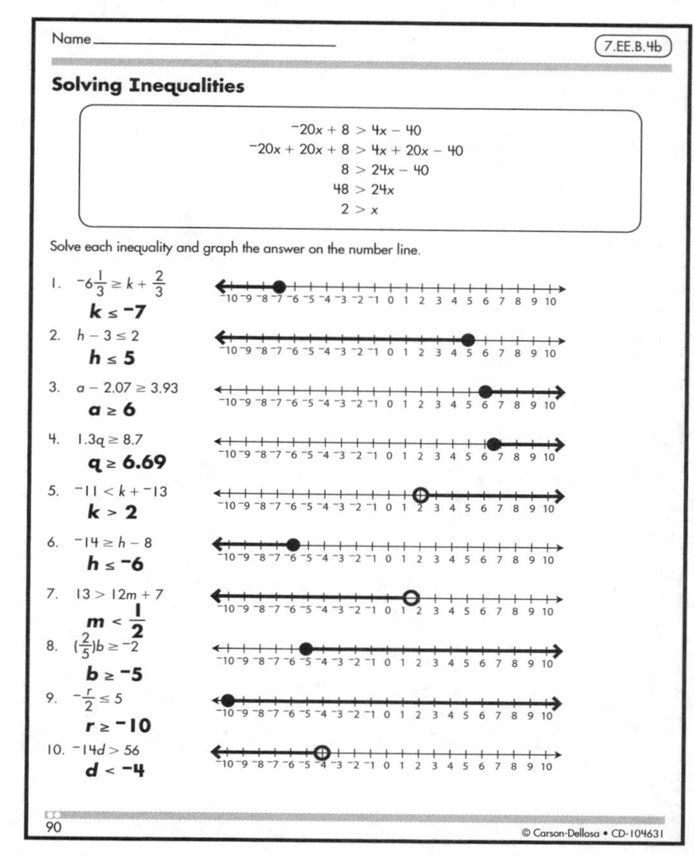

Name____________________ (7.EE.B.4b)

Solving Inequalities

$-20x + 8 > 4x - 40$
$-20x + 20x + 8 > 4x + 20x - 40$
$8 > 24x - 40$
$48 > 24x$
$2 > x$

Solve each inequality and graph the answer on the number line.

1. $-6\frac{1}{3} \geq k + \frac{2}{3}$ — **$k \leq -7$**
2. $h - 3 \leq 2$ — **$h \leq 5$**
3. $a - 2.07 \geq 3.93$ — **$a \geq 6$**
4. $1.3q \geq 8.7$ — **$q \geq 6.69$**
5. $-11 < k + -13$ — **$k > 2$**
6. $-14 \geq h - 8$ — **$h \leq -6$**
7. $13 > 12m + 7$ — **$m < \frac{1}{2}$**
8. $(\frac{2}{5})b \geq -2$ — **$b \geq -5$**
9. $-\frac{r}{2} \leq 5$ — **$r \geq -10$**
10. $-14d > 56$ — **$d < -4$**

90 © Carson-Dellosa • CD-104631

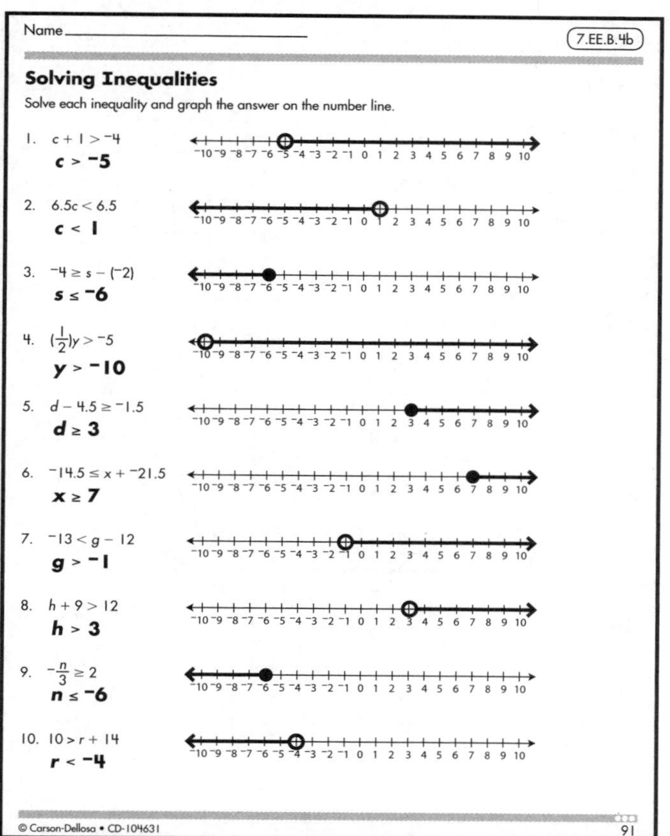

Name____________________ (7.EE.B.4b)

Solving Inequalities

Solve each inequality and graph the answer on the number line.

1. $c + 1 > -4$ — **$c > -5$**
2. $6.5c < 6.5$ — **$c < 1$**
3. $-4 \geq s - (-2)$ — **$s \leq -6$**
4. $(\frac{1}{2})y > -5$ — **$y > -10$**
5. $d - 4.5 \geq -1.5$ — **$d \geq 3$**
6. $-14.5 \leq x + -21.5$ — **$x \geq 7$**
7. $-13 < g - 12$ — **$g > -1$**
8. $h + 9 > 12$ — **$h > 3$**
9. $-\frac{n}{3} \geq 2$ — **$n \leq -6$**
10. $10 > r + 14$ — **$r < -4$**

© Carson-Dellosa • CD-104631 91

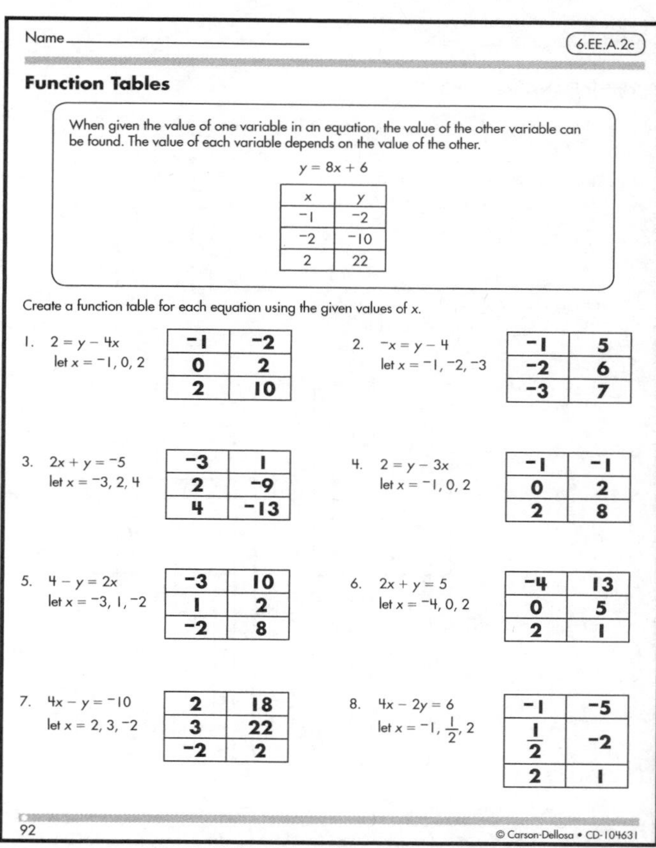

Name____________________ (6.EE.A.2c)

Function Tables

When given the value of one variable in an equation, the value of the other variable can be found. The value of each variable depends on the value of the other.

$y = 8x + 6$

x	y
−1	−2
−2	−10
2	22

Create a function table for each equation using the given values of x.

1. $2 = y - 4x$, let $x = -1, 0, 2$

−1	**−2**
0	**2**
2	**10**

2. $-x = y - 4$, let $x = -1, -2, -3$

−1	**5**
−2	**6**
−3	**7**

3. $2x + y = -5$, let $x = -3, 2, 4$

−3	**1**
2	**−9**
4	**−13**

4. $2 = y - 3x$, let $x = -1, 0, 2$

−1	**−1**
0	**2**
2	**8**

5. $4 - y = 2x$, let $x = -3, 1, -2$

−3	**10**
1	**2**
−2	**8**

6. $2x + y = 5$, let $x = -4, 0, 2$

−4	**13**
0	**5**
2	**1**

7. $4x - y = -10$, let $x = 2, 3, -2$

2	**18**
3	**22**
−2	**2**

8. $4x - 2y = 6$, let $x = -1, \frac{1}{2}, 2$

−1	**−5**
$\frac{1}{2}$	**−2**
2	**1**

92 © Carson-Dellosa • CD-104631

Answer Key

Name ______________________ (6.EE.A.2c)

Function Tables

Remember, when given the value of one variable in an equation, the value of the other variable can be found. The value of each variable depends on the value of the other.

Create a function table for each equation using the given values of x.

1. $y + 3x = 7$, let $x = {}^-9, {}^-5, 8$

x	y
−9	34
−5	22
8	−17

2. $y = {}^-2x + 8$, let $x = {}^-7, {}^-1, 2$

x	y
−7	22
−1	10
2	4

3. ${}^-7 - 6x = y$, let $x = {}^-6, {}^-4, 9$

x	y
−6	29
−4	17
9	−61

4. $y + 2 = {}^-5x$, let $x = {}^-4, {}^-1, 3$

x	y
−4	18
−1	3
3	−17

5. $y = {}^-4x - 6$, let $x = {}^-6, {}^-3, 0$

x	y
−6	18
−3	6
0	−6

6. $y - 9x = {}^-5$, let $x = {}^-7, 3, 7$

x	y
−7	−68
3	22
7	58

7. $6 = y - 8x$, let $x = {}^-5, 1, 8$

x	y
−5	−34
1	14
8	70

8. $y + 4 = {}^-9x$, let $x = {}^-2, 0, 2$

x	y
−2	14
0	−4
2	−22

9. $y = 5x + 3$, let $x = {}^-2, 2, 4$

x	y
−2	−7
2	13
4	23

10. $2x = y - 2$, let $x = {}^-3, 3, 6$

x	y
−3	−4
3	8
6	14

 93

Name ______________________ (6.EE.A.2c)

Function Tables

Create a function table for each equation using the given values of x.

1. $y = {}^-2x - 3$, let $x = {}^-8, 1, 6$

x	y
−8	13
1	−5
6	−15

2. $y + \frac{1}{2}x = 3$, let $x = {}^-5, 1, 5$

x	y
−5	$5\frac{1}{2}$
1	$2\frac{1}{2}$
5	$\frac{1}{2}$

3. $y + 1 = \frac{1}{2}x$, let $x = {}^-7, 0, 7$

x	y
−7	$-4\frac{1}{2}$
0	−1
7	$2\frac{1}{2}$

4. $y + \frac{1}{4}x = {}^-5$, let $x = {}^-8, {}^-2, 4$

x	y
−8	−3
−2	$-4\frac{1}{2}$
4	−6

5. $y = {}^-3x - 4$, let $x = {}^-4, 0, 3$

x	y
−4	8
0	−4
3	−13

6. $\frac{1}{2}x - 2 = y$, let $x = {}^-7, {}^-4, 4$

x	y
−7	$-5\frac{1}{2}$
−4	−4
4	0

7. $5 + y = -\frac{1}{2}x$, let $x = {}^-2, 6, 8$

x	y
−2	−4
6	−8
8	−9

8. $y = \frac{1}{4}x + 2$, let $x = {}^-7, 0, 8$

x	y
−7	$\frac{1}{4}$
0	2
8	4

9. $1 = -\frac{1}{2}x - y$, let $x = {}^-6, 2, 3$

x	y
−6	2
2	−2
3	$-2\frac{1}{2}$

10. $y = -\frac{1}{4}x + 2$, let $x = {}^-6, {}^-1, 3$

x	y
−6	$3\frac{1}{2}$
−1	$2\frac{1}{4}$
3	$1\frac{1}{4}$

11. $y + \frac{1}{4}x = {}^-2$, let $x = {}^-7, 0, 7$

x	y
−7	$-\frac{1}{4}$
0	−2
7	$-3\frac{3}{4}$

12. $y + 2 = \frac{1}{4}x$, let $x = {}^-8, {}^-5, {}^-4$

x	y
−8	−4
−5	$-3\frac{1}{4}$
−4	−3

94

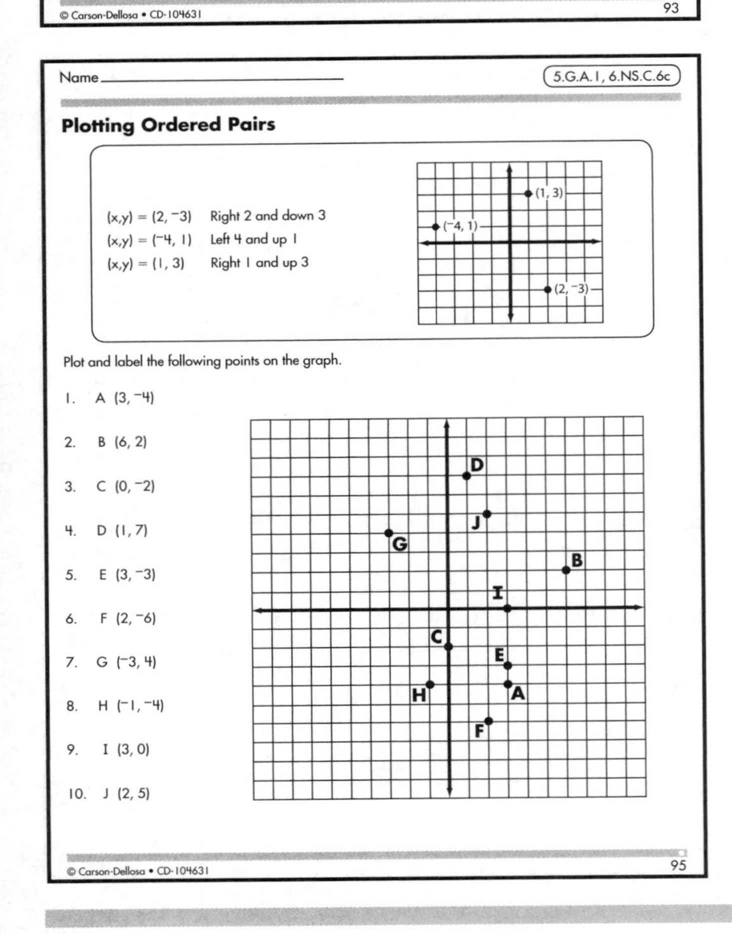
Name ______________________ (5.G.A.1, 6.NS.C.6c)

Plotting Ordered Pairs

(x,y) = (2, −3) Right 2 and down 3
(x,y) = (−4, 1) Left 4 and up 1
(x,y) = (1, 3) Right 1 and up 3

Plot and label the following points on the graph.

1. A (3, −4)
2. B (6, 2)
3. C (0, −2)
4. D (1, 7)
5. E (3, −3)
6. F (2, −6)
7. G (−3, 4)
8. H (−1, −4)
9. I (3, 0)
10. J (2, 5)

 95

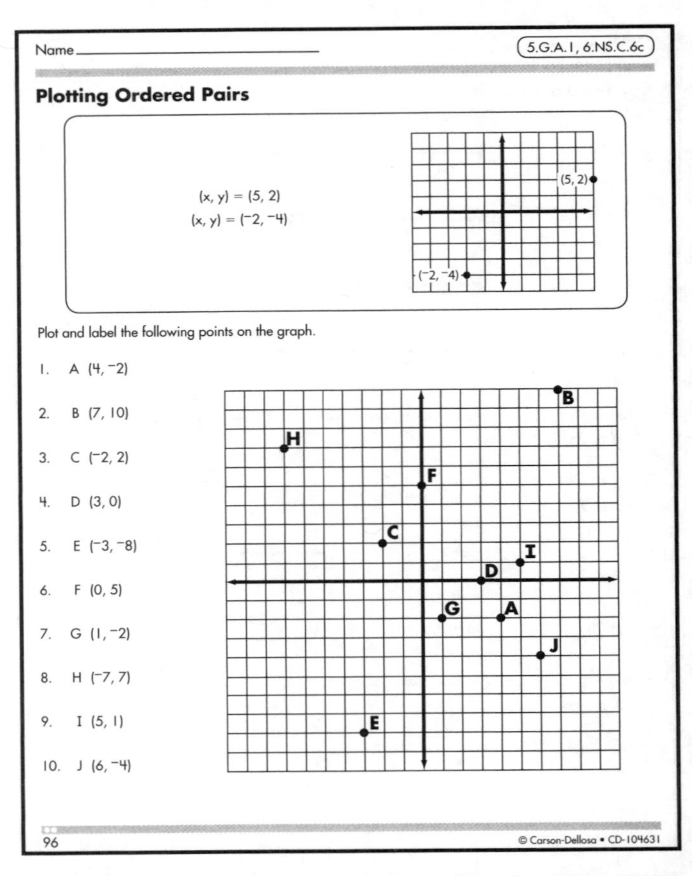
Name ______________________ (5.G.A.1, 6.NS.C.6c)

Plotting Ordered Pairs

(x, y) = (5, 2)
(x, y) = (−2, −4)

Plot and label the following points on the graph.

1. A (4, −2)
2. B (7, 10)
3. C (−2, 2)
4. D (3, 0)
5. E (−3, −8)
6. F (0, 5)
7. G (1, −2)
8. H (−7, 7)
9. I (5, 1)
10. J (6, −4)

96

Answer Key

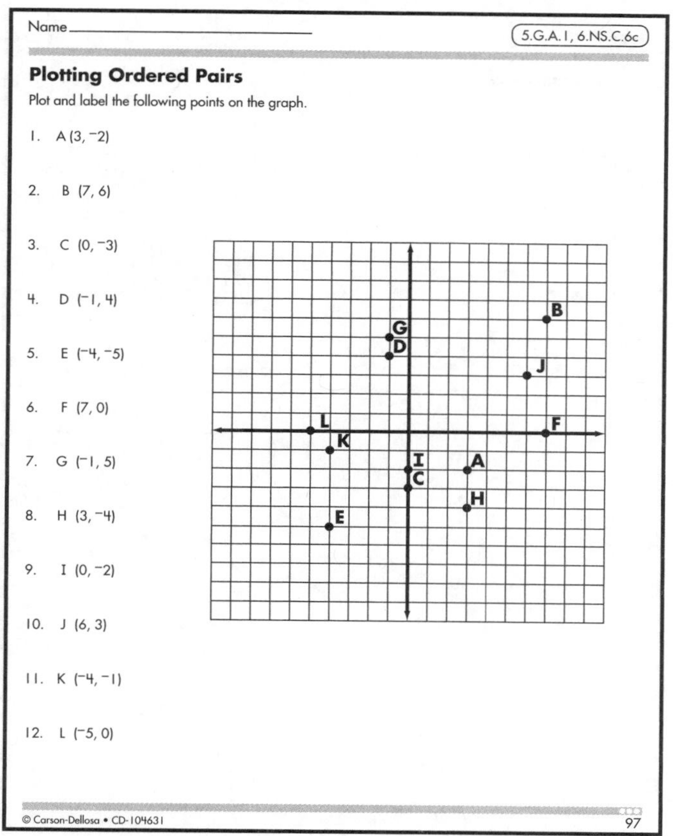

Name ____________________ (5.G.A.1, 6.NS.C.6c)

Plotting Ordered Pairs

Plot and label the following points on the graph.

1. A (3, ⁻2)
2. B (7, 6)
3. C (0, ⁻3)
4. D (⁻1, 4)
5. E (⁻4, ⁻5)
6. F (7, 0)
7. G (⁻1, 5)
8. H (3, ⁻4)
9. I (0, ⁻2)
10. J (6, 3)
11. K (⁻4, ⁻1)
12. L (⁻5, 0)

© Carson-Dellosa • CD-104631 97

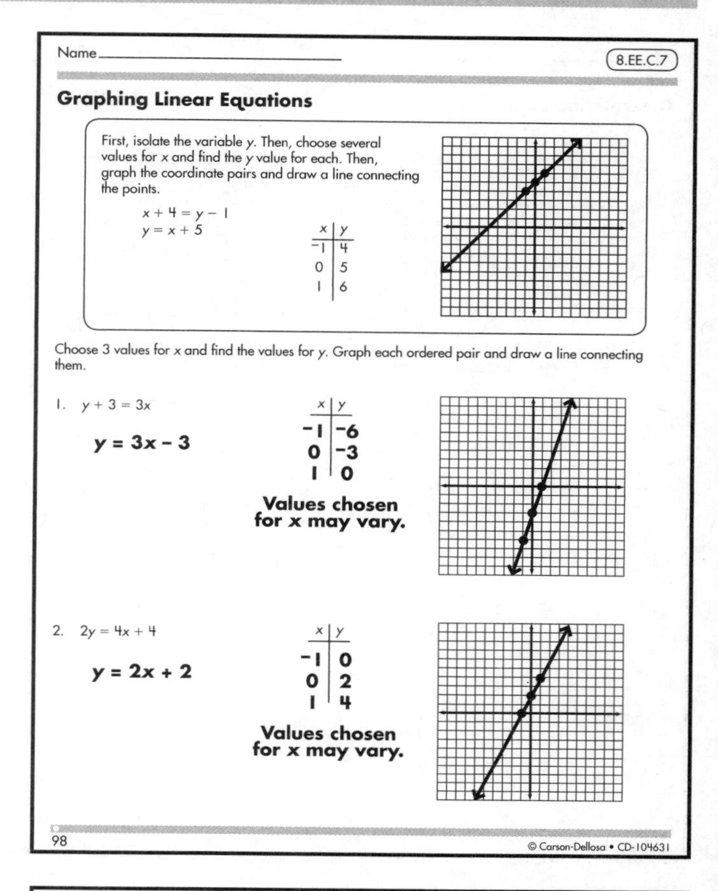

Name ____________________ (8.EE.C.7)

Graphing Linear Equations

First, isolate the variable *y*. Then, choose several values for *x* and find the *y* value for each. Then, graph the coordinate pairs and draw a line connecting the points.

$x + 4 = y - 1$
$y = x + 5$

x	y
⁻1	4
0	5
1	6

Choose 3 values for *x* and find the values for *y*. Graph each ordered pair and draw a line connecting them.

1. $y + 3 = 3x$

y = 3x − 3

x	y
-1	-6
0	-3
1	0

Values chosen for x may vary.

2. $2y = 4x + 4$

y = 2x + 2

x	y
-1	0
0	2
1	4

Values chosen for x may vary.

98 © Carson-Dellosa • CD-104631

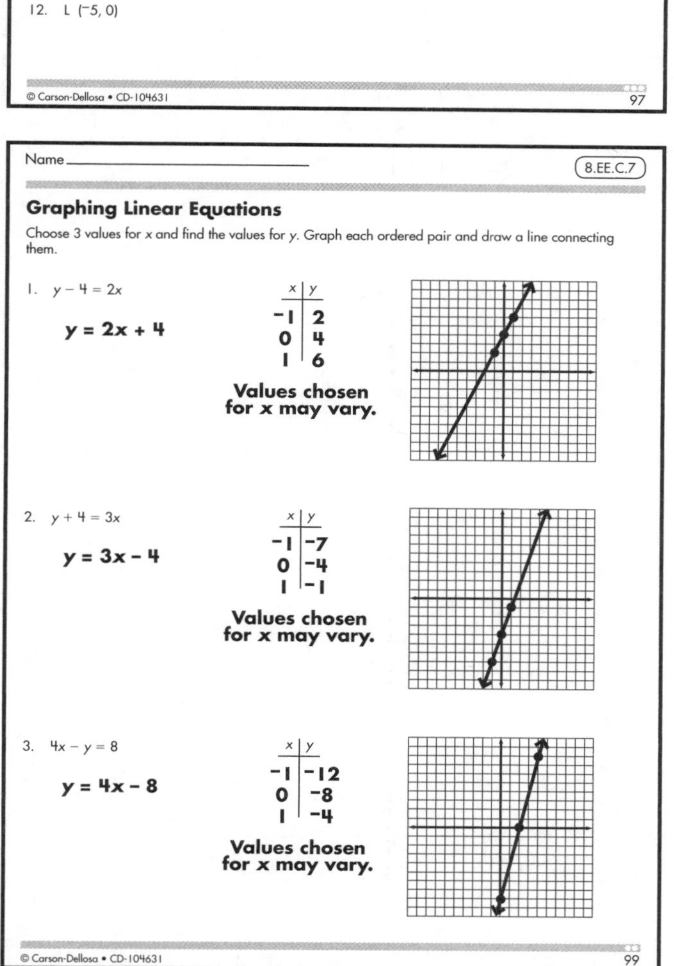

Name ____________________ (8.EE.C.7)

Graphing Linear Equations

Choose 3 values for *x* and find the values for *y*. Graph each ordered pair and draw a line connecting them.

1. $y - 4 = 2x$

y = 2x + 4

x	y
-1	2
0	4
1	6

Values chosen for x may vary.

2. $y + 4 = 3x$

y = 3x − 4

x	y
-1	-7
0	-4
1	-1

Values chosen for x may vary.

3. $4x - y = 8$

y = 4x − 8

x	y
-1	-12
0	-8
1	-4

Values chosen for x may vary.

© Carson-Dellosa • CD-104631 99

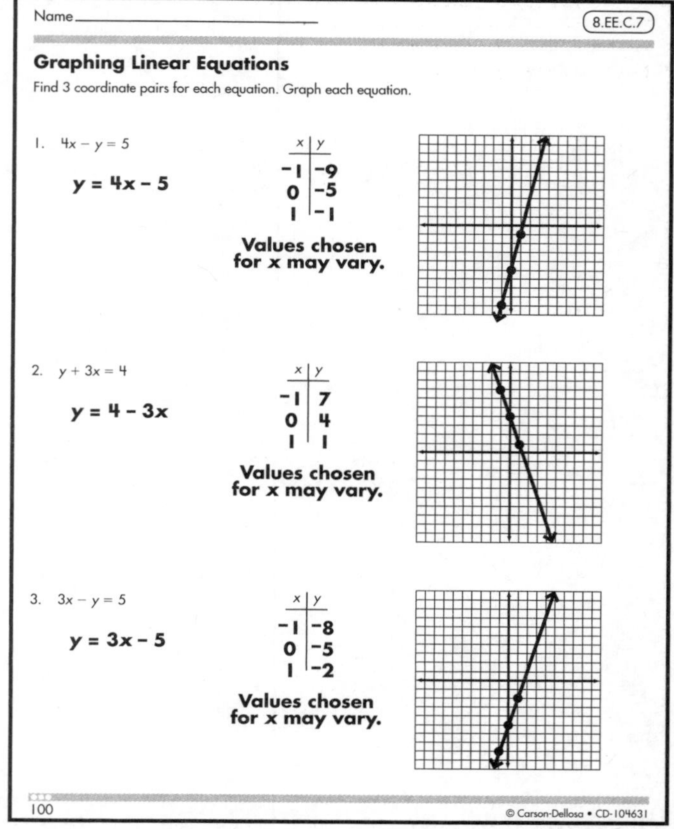

Name ____________________ (8.EE.C.7)

Graphing Linear Equations

Find 3 coordinate pairs for each equation. Graph each equation.

1. $4x - y = 5$

y = 4x − 5

x	y
-1	-9
0	-5
1	-1

Values chosen for x may vary.

2. $y + 3x = 4$

y = 4 − 3x

x	y
-1	7
0	4
1	1

Values chosen for x may vary.

3. $3x - y = 5$

y = 3x − 5

x	y
-1	-8
0	-5
1	-2

Values chosen for x may vary.

100 © Carson-Dellosa • CD-104631

Answer Key

Name ______________________ 8.EE.B.5, 8.F.B.4

Determining Slope

The **slope** of a line describes how steep it is. It is calculated by dividing the vertical rate of change (change in *y*) by the horizontal change (change in *x*). You can subtract the change in either order (A − B, or A − B), as long as both are subtracted in the same order.

A(⁻1, 4) B(3, 2)

slope (m) = $2 - 4 \div 3 - (^{-}1) = \frac{2-4}{3-(^{-}1)}$

$m = {}^{-}\frac{1}{4}$

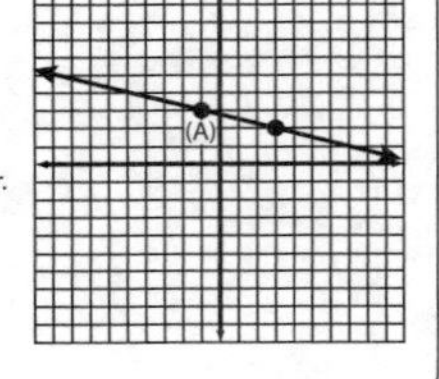

Find the slope of the line that contains each pair of points.

1. A(2, 4), B(4, 6) **m = 1**
2. A(3, 2), B(⁻2, ⁻8) **m = 2**
3. A(⁻1, 3), B(⁻2, 5) **m = ⁻2**
4. A(8, ⁻3), B(10, 0) **m = $\frac{3}{2}$**
5. A(0, 0), B(6, ⁻3) **m = ⁻$\frac{1}{2}$**
6. A(3, 4), B(⁻1, 4) **m = 0**
7. A(⁻4, 7), B(⁻7, 15) **m = ⁻$\frac{8}{3}$**
8. A(2, ⁻2), B(6, 5) **m = $\frac{7}{4}$**

© Carson-Dellosa • CD-104631 101

Name ______________________ 8.EE.B.5, 8.F.B.4

Determining Slope

Remember, find the slope by dividing the vertical rate of change (change in *y*) by the horizontal change (change in *x*).

A(2, 4) B(4, 9)

slope (m) = $4 - 9 \div 2 - 4 \quad \frac{4-9}{2-4}$

$m = \frac{^{-}5}{^{-}2} = \frac{5}{2}$

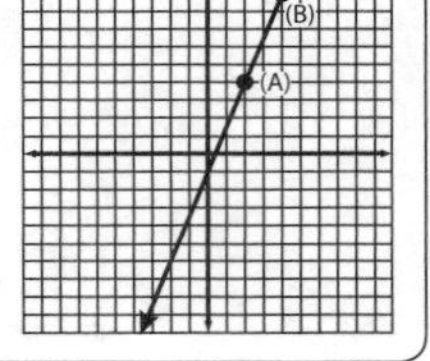

Find the slope of the line that contains each pair of points.

1. A(0, 1), B(2, 4) **m = $\frac{3}{2}$**
2. A(2, 2), B(3, 3) **m = 1**
3. A(2, 0), B(3, 0) **m = 0**
4. A(3, 0), B(2, 3) **m = ⁻3**
5. A(⁻1, 0), B(⁻3, 1) **m = ⁻$\frac{1}{2}$**
6. A(⁻1, ⁻1), B(⁻3, 2) **m = ⁻$\frac{3}{2}$**
7. A(1, 4), B(⁻1, 2) **m = 1**
8. A(⁻4, ⁻4), B(0, ⁻3) **m = $\frac{1}{4}$**
9. A(⁻1, 2), B(⁻3, ⁻3) **m = $\frac{5}{2}$**
10. A(0, 3), B(3, 4) **m = $\frac{1}{3}$**

102 © Carson-Dellosa • CD-104631

Name ______________________ 8.EE.B.5, 8.F.B.4

Determining Slope

Find the slope of the line that contains each pair of points.

1. A(0, ⁻2), B(1, 3) **m = 5**
2. A(2, ⁻1), B(5, ⁻2) **m = ⁻$\frac{1}{3}$**
3. A(⁻3, 0), B(⁻2, 5) **m = 5**
4. A(⁻4, ⁻1), B(1, 3) **m = $\frac{4}{5}$**
5. A(0, 1), B(⁻4, 4) **m = ⁻$\frac{3}{4}$**
6. A(1, 0), B(4, 1) **m = $\frac{1}{3}$**
7. A(⁻2, 0), B(0, ⁻2) **m = ⁻1**
8. A(⁻1, ⁻2), B(1, 4) **m = 3**
9. A(⁻1, 2), B(⁻2, 5) **m = ⁻3**
10. A(0, ⁻1), B(5, ⁻2) **m = ⁻$\frac{1}{5}$**
11. A(⁻3, 1), B(2, ⁻2) **m = ⁻$\frac{3}{5}$**
12. A(10, 3), B(7, 9) **m = ⁻2**

© Carson-Dellosa • CD-104631 103

Congratulations!

____

receives this award for

____

Signed ______________________________ **Date** ________________

Write as a decimal.

$$\frac{2}{3}$$

Write as a decimal.

$$\frac{1}{12}$$

Write as a decimal.

$$\frac{1}{5}$$

Write as a decimal.

$$\frac{1}{4}$$

Write as a decimal.

$$\frac{3}{10}$$

Write as a decimal.

$$\frac{33}{100}$$

Write as a decimal.

$$\frac{23}{40}$$

Write as a decimal.

$$\frac{7}{1000}$$

Write as a fraction.

0.45

Write as a fraction.

0.005

Write as a fraction.

0.25

Write as a fraction.

0.15

Write as a fraction.

0.05

Write as a fraction.

0.2

Write as a fraction.

0.125

Write as a fraction.

0.7

0.25	0.2	$0.08\overline{3}$	$0.\overline{6}$
0.007	0.575	0.33	0.3
$\frac{3}{20}$	$\frac{1}{4}$	$\frac{1}{200}$	$\frac{9}{20}$
$\frac{7}{10}$	$\frac{1}{8}$	$\frac{1}{5}$	$\frac{1}{20}$

Round to the nearest whole number.

45.21

Round to the nearest whole number.

356.4

Round to the nearest whole number.

4,515.321

Round to the nearest whole number.

901.14

Round to the nearest ten.

341.5

Round to the nearest ten.

3,357

Round to the nearest hundred.

304,313

Round to the nearest hundred.

42,555.6

Round to the nearest tenth.

32.54

Round to the nearest tenth.

201.15

Round to the nearest tenth.

2,343.38

Round to the nearest tenth.

4.23

Round to the nearest hundredth.

5.455

Round to the nearest hundredth.

3,295.675

Round to the nearest hundredth.

203.421

Round to the nearest hundredth.

12.901

901	4,515	356	45
42,600	304,300	3,360	340
4.2	2,343.4	201.2	32.5
12.90	203.42	3,295.68	5.46

$(-7)(-2)$	$-4 \times (-2)$	$49 \div (-7)$	$-36 \div 9$
$(-4) \cdot (4)$	$-63 \div (-6)$	$-44 \div (-11)$	$4 \times (-5)$
$56 \div (-8)$	$(-7)(5)$	$24 \div (-6)$	$-15 \div 5$
$-12 \cdot 6$	$(-11)(10)$	$-36 \div (-12)$	$(5)(-12)$

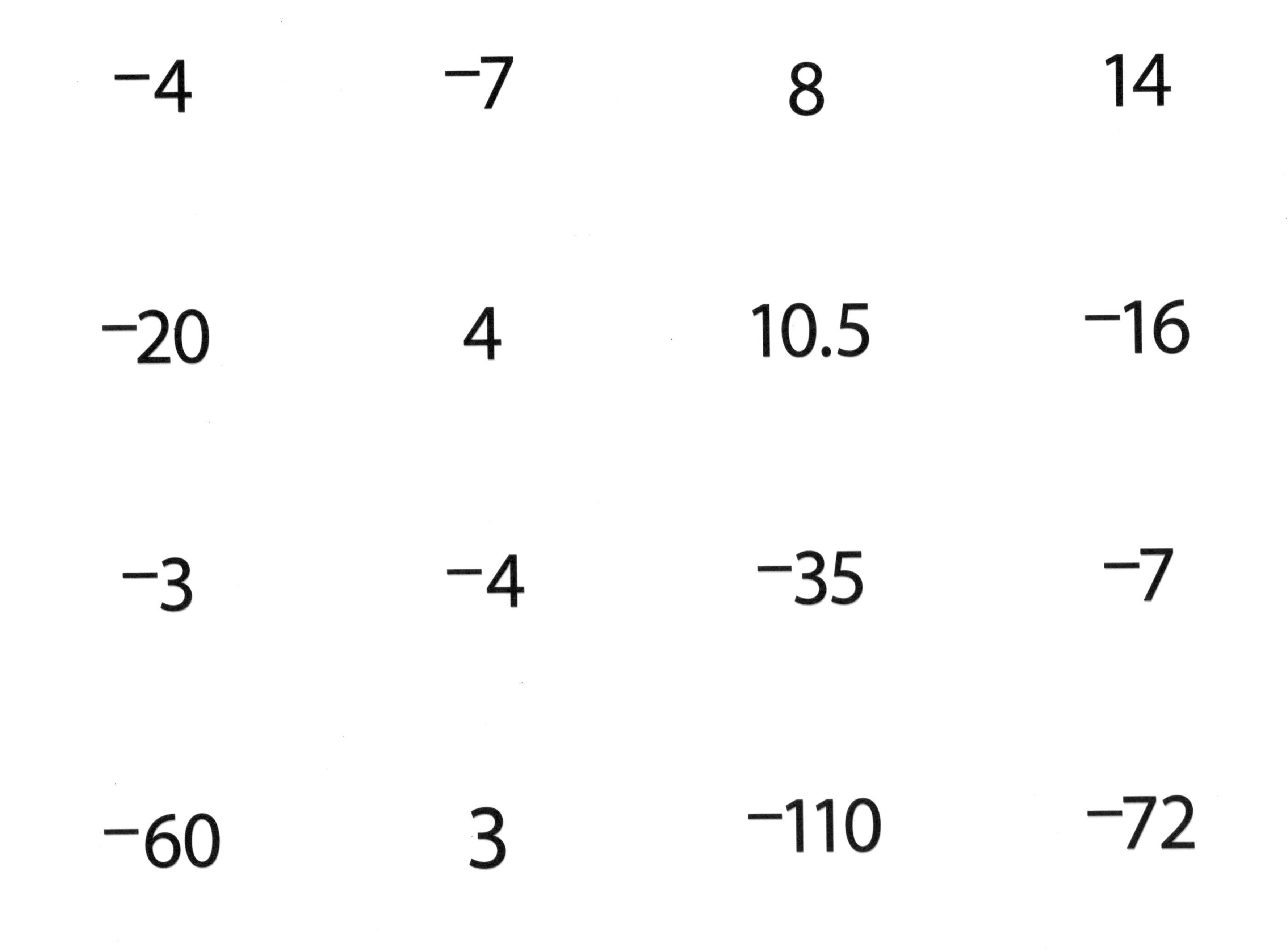
$^{-}4$
$^{-}7$
8
14
$^{-}20$
4
10.5
$^{-}16$
$^{-}3$
$^{-}4$
$^{-}35$
$^{-}7$
$^{-}60$
3
$^{-}110$
$^{-}72$

$5 + 3 \times 4 =$	$14 \div 2 \times 6 =$	$11 - 3 + 5 =$	$3(5 + 4) =$
$24 \div 6 - 2 =$	$8 + 6 \div 3 =$	$3 \cdot 5 - 4 \cdot 8 =$	$6 - 49 \div 7 =$
$5 + 3(20 \div 4) =$	$35 \div 5 + 3 =$	$7 + 2 \cdot 2 =$	$21 \div (2 + 5) =$
$3^2 - 2 \cdot 3 =$	$(4 + 1)^2 =$	$15 \div 5 \times 2 =$	$81 \div 3^2 - 2 =$

27	13	42	17
$^{-}1$	$^{-}17$	10	2
3	11	10	20
7	6	25	3

$^{-}8 + 14 =$	$^{-}12 + 8 =$	$^{-}3 - 5 =$	$3 - 7 =$
$^{-}4 - (^{-}2) =$	$^{-}5 + (^{-}9) =$	$7 + (^{-}6) =$	$^{-}5 + (^{-}5) =$
$15 + (^{-}7) =$	$^{-}19 - (^{-}4) =$	$4 - 8 =$	$16 - 8 =$
$^{-}13 + 7 =$	$^{-}2 + (^{-}6) =$	$^{-}5 + 8 =$	$4 - (^{-}3) =$

$^{-}4$	$^{-}8$	$^{-}4$	6
$^{-}10$	1	$^{-}14$	$^{-}2$
8	$^{-}4$	$^{-}15$	8
7	3	$^{-}8$	$^{-}6$

$(^-3) - 5 =$	$2 - 9 =$	$(^-6) - (^-6) =$	$2 + 6 =$
$^-12 - (^-4) =$	$5 - 10 =$	$7 - (^-3) =$	$^-11 - (^-5) =$
$^-4 + 5 =$	$^-3 - 3 =$	$^-4 + (^-3) =$	$3 - 5 =$
$7 - (^-4) =$	$7 - 13 =$	$^-4 - (^-8) =$	$^-3 - (^-4) =$

8	0	$^{-}7$	$^{-}8$
$^{-}6$	10	$^{-}5$	$^{-}8$
$^{-}2$	$^{-}7$	$^{-}6$	1
1	4	$^{-}6$	11